Natural Computing Series

For further volumes:
http://www.springer.com/series/4190

Daniel S. Yeung · Ian Cloete · Daming Shi ·
Wing W.Y. Ng

Sensitivity Analysis for Neural Networks

Authors

Prof. Daniel S. Yeung
School of Computer Science
and Engineering
South China University of Technology
Wushan Rd.
TianHe District
Guangzhou, China
danyeung@ieee.org

Prof. Daming Shi
School of Electrical Engineering and Computer
Science
Kyungpook National University
Buk-gu, Daegu
South Korea
asdmshi@ntu.edu.sg

Prof. Ian Cloete
President
Campus 3
International University in Germany
76646 Bruchsal, Germany
ian.cloete@i-u.de; president@i-u.de

Dr. Wing W.Y. Ng
School of Computer Science
and Engineering
South China University of Technology
Wushan Rd.
TianHe District
Guangzhou, China
wingng@ieee.org

ISSN 1619-7127
ISBN 978-3-642-26139-8 e-ISBN 978-3-642-02532-7
DOI 10.1007/978-3-642-02532-7
Springer Heidelberg Dordrecht London New York

ACM Computing Classification (1998): I.2.6, F.1.1

Cover design: KuenkelLopka GmbH

Printed on acid-free paper

Springer is part of Springer Science+Business Media (www.springer.com)

Preface

Neural networks provide a way to realize one of our human dreams to make machines think like us. Artificial neural networks have been developed since Rosenblatt proposed the Perceptron in 1958. Today, many neural networks are not treated as black boxes any more. Issues such as robustness and generalization abilities have been brought to the fore. The advances in neural networks have led to more and more practical applications in pattern recognition, financial engineering, automatic control and medical diagnosis, to name just a few.

Sensitivity analysis dates back to the 1960s, when Widrow investigated the probability of misclassification due to weight perturbations, which are caused by machine imprecision and noisy input. For the purpose of analysis, these perturbations can be simulated by embedding disturbance into the original inputs or connection weights. The initial idea of sensitivity analysis was then extended to optimization and to applications of neural networks, such as sample reduction, feature selection, active learning and critical vector learning.

This text should primarily be of interest to graduate students, academics, and researchers in branches of neural networks, artificial intelligence, machine learning, applied mathematics and computer engineering where sensitivity analysis of neural networks and related concepts are used. We have made an effort to make the book accessible to such a cross-disciplinary audience.

The book is organized into eight chapters, of which Chap. 1 gives an introduction to the various neural network structures and learning schemes. A literature review on the methodologies of sensitivity analysis is presented in Chap. 2. Different from the traditional hypersphere model, the hyper-rectangle model described in Chap. 3 is especially suitable for the most popular and general feedforward network: the multilayer Perceptron. In Chap. 4, the activation function is also involved in the calculation of the sensitivity analysis by parameterizing. The sensitivity analysis of radial basis function networks is discussed in Chaps. 5 and 6, with the former giving a generalization error model whereas the latter concerns optimizing the hidden neurons. In Chap. 7, sensitivity is measured in order to encode prior knowledge into a neural network. In Chap. 8, sensitivity analysis is applied in many applications, such as dimensionality reduction, network optimization and selective learning.

We would like to express our thanks to many colleagues, friends and students who provided reviews of different chapters of this manuscript. They include Minh

Nhut Nguyen, Xiaoqin Zeng, Patrick Chan, Xizhao Wang, Fei Chen and Lu He. We often find ourselves struggling with many competing demands for our time and effort. As a result, our families, especially our beloved spouses, are the ones who suffer the most. We are delighted to dedicate this work to Foo-Lau Yeung, Wilma Cloete and Jian Liu.

It is with great humility that we would like to acknowledge our Good Lord as the true creator of all knowledge. This work is the result of our borrowing a small piece of knowledge from Him.

8 July 2009

Daniel S. Yeung
Ian Cloete
Daming Shi
Wing W.Y. Ng

Contents

Chapter 1
Introduction to Neural Networks

The human brain consists of ten billion densely interconnected nerve cells, called *neurons*; each connected to about 10,000 other neurons, with 60 trillion connections, *synapses*, between them. By using multiple neurons simultaneously, the brain can perform its functions much faster than the fastest computers in existence today. On the other hand, a neuron can be considered as a basic information-processing unit, whereas our brain can be considered as a highly complex, nonlinear and parallel biological information-processing network, in which information is stored and processed simultaneously. Learning is a fundamental and essential characteristic of biological neural networks. The ease with which they can learn led to attempts to emulate a biological neural network in a computer.

In the 1940s, McCulloch and Pitts proposed a model for biological neurons and biological neural networks. A stimulus is transmitted from dendrites to a soma via synapses, and axons transmit the response of one soma to another, as shown in Fig. 1.1. Inspired by the mechanism for learning in biological neurons, artificial neurons and artificial neural networks can perform arithmetic functions, with cells corresponding to neurons, activations corresponding to neuronal firing rates, connections corresponding to synapses, and connection weights corresponding to synaptic strengths, as shown in Fig. 1.1. The analogy between biological neurons and artificial neurons is made in Table 1.1. However, neural networks are far too simple to serve as realistic brain models on the cell level, but they might serve as very good models for the essential information processing tasks that organisms perform. This remains an open question because we have so little understanding of how the brain actually works (Gallant, 1993).

In a neural network, *neurons* are joined by directed arcs – *connections*. The neurons and arcs constitute the network *topology*. Each arc has a numerical *weight* that specifies the influence between two neurons. Positive weights indicate reinforcement; negative weights represent inhibition. The weights determine the behavior of the network, playing somewhat the same role as in a conventional computer program. Typically, there are many *inputs* for a single neuron, and a subsequent *output* of an *activation function* (or *transfer function*). Some frequently used activation functions include:

D.S. Yeung et al., *Sensitivity Analysis for Neural Networks*, Natural Computing Series, DOI 10.1007/978-3-642-02532-7_1, © Springer-Verlag Berlin Heidelberg 2010

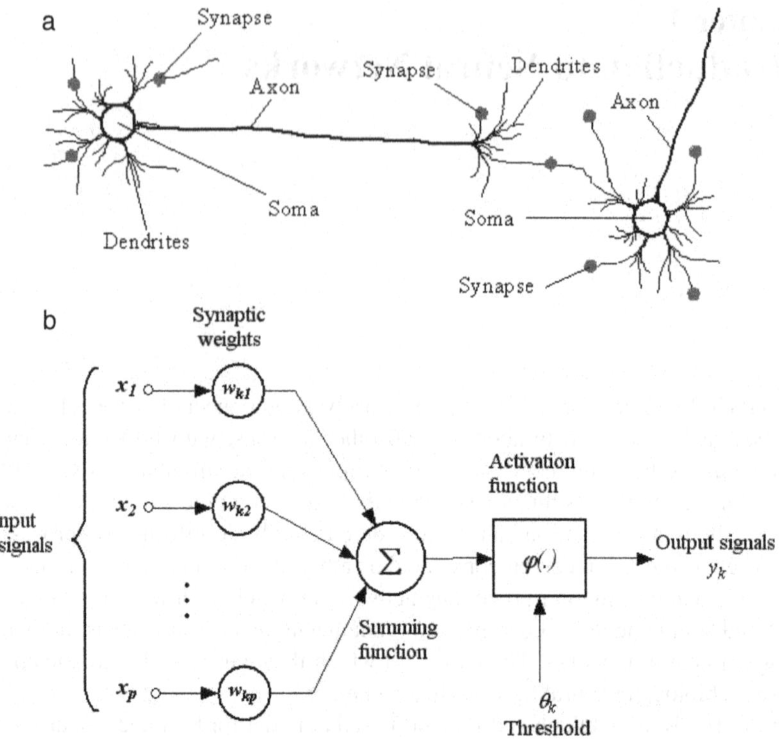

Fig. 1.1 Biological motivations of neural networks. (a) Neuroanatomy of living animals. (b) Connections of an artificial neuron

- Linear Function: $\Phi(u) = u$
- Log-Sigmoid Function: $\Phi(u) = \dfrac{1}{1+e^{-u}}$
- Hard Limit function: $\Phi(u) = \begin{cases} 1 \text{ if } u \geq 0 \\ 0 \text{ otherwise} \end{cases}$

Rosenblatt (1958) devised the Perceptron, which is now widely used. Widrow and Hoff (1960) proposed the Adaline at the same time. The Perceptron and Adaline were the first practical networks and learning rules that demonstrated the high potential of neural networks.

Table 1.1 Analogy between biological and artificial neurons

Biological Neurons	Artificial Neurons
Soma	Sum + Activation Function
Dendrite	Input
Axon	Output
Synapse	Weight

Minsky and Papert (1969) criticized the Perceptron, because they found it is not powerful enough to do tasks such as parity and connectedness. As long as a neural network consists of a *linear combiner* followed by a *nonlinear element*, a single-layer Perceptron can perform pattern classification *only* on *linearly separable* patterns, regardless of the form of nonlinearity used. Minsky and Papert demonstrated the limitations of the Perceptron with the simplest XOR problem, and as a consequence the research in the field was suspended due to the lack of funding. Because of Minsky and Papert's conclusion, people lost confidence in neural networks. In the 1970s, the progress on neural networks continued at a much reduced pace, although some researchers, such as Amari, Anderson, Fukushima, Grossberg and Kohonen, kept working on it.

In 1985, Ackley, Hinton and Sejnowski described the Boltzmann machine, which is more powerful than a Perceptron, and demonstrated a successful application – NETtalk.

Rumelhart, Hinton and Williams (1986) are recognized for their milestone work – the Multilayer Perceptron (MLP) with backpropagation, which remained the dominant neural network architecture for more than ten years. A sufficiently big MLP has been proven to be able to learn any function, and many variants of MLP have been proposed since then. In 1988, Radial Basis Functions (RBFs), introduced as an alternative to MLPs, speeded up MLP training (fast-prop).

From the 1990s to date, more and more research has been done to improve neural networks. The research agenda includes regularizers, probabilistic (Bayesian) inference, structural risk minimization and incorporation with evolutionary computation.

1.1 Properties of Neural Networks

Neural networks can be described according to their *network, cell, dynamic*, and *learning properties* as follows:

Network Properties. A neural network is an architecture consisting of many neurons, which work together to respond to the inputs. We sometimes consider a network as a black-box function. Here the external world presents inputs to the input cells and receives network outputs from the output cells. Intermediate cells are not seen externally, and for this reason they are often called *hidden units*. We classify networks as either *feedforward networks* if they do not contain directed cycles or *recurrent networks* if they do contain such cycles. It is often convenient to organize the cells of a network into layers.

Cell Properties. Each cell computes a single (numerical) cell output or *activation*. Cell inputs and activations may be *discrete*, taking values $\{0, 1\}$ or $\{-1, 0, 1\}$, or they may be *continuous*, assuming values in the interval $[0, 1]$ or $[-1, +1]$. Typically, every cell uses the same algorithm for computing its activation. The activation for a cell must be computed from the activations of cells directly connected to it and the corresponding weights for those connections. Every cell (except for input cells) computes its new activation as a function of the weighted sum of the inputs from directly connected cells.

Dynamic Properties. A neural network model must specify the time when each cell computes its new activation value and when the change to that cell's output actually takes place. Usually in feedforward models, cells are visited in a fixed order, and each cell reevaluates and changes its activation before the next one is visited. In this case the network achieves a steady state after one pass through the cells, provided the cells are correctly numbered. For recurrent models there are several possibilities. One possibility is to make one ordered pass through the cells, as with feedforward models; but we are no longer guaranteed that the network will reach steady state. Another possibility is to make repeated passes through the network. For discrete models the network will either reach a steady state or it will cycle; for continuous models the network will either reach a steady state, cycle, approach a steady state in the limit, blow up, or some combination of these things. A third possibility is to compute new activations for all cells simultaneously and then make changes to the cell outputs simultaneously. This is similar to the previous case. Still another possibility is to select a cell at random, compute its new activation, and then change its output before selecting the next cell. In this case we have no guarantee of any sort of limiting or cyclic behavior unless other constraints are placed upon the model.

Learning Properties. Each neural network needs to be trained to respond to the inputs, so it must be associated with one or more algorithms for *machine learning*. Machine learning refers to computer models that improve neural network performance in significant ways based upon the input data. Machine learning techniques are usually divided into supervised and unsupervised models. In *supervised learning* a *teacher* or *critic* supplies additional input data that gives a measure of how well a program is performing during a training phase. The most common form of supervised learning is trying to duplicate behavior specified by a finite set of *training examples*, where each example consists of input data and the corresponding correct output. In *unsupervised learning* there is no performance evaluation available. Without any specific knowledge of what constitutes a correct answer and what constitutes an incorrect answer, the most that can be expected from these models is the construction of groups of similar input patterns. In the pattern recognition literature this is known as *clustering*.

Neural networks have potential as intelligent control systems because they can learn and adapt, they can approximate nonlinear functions, they are suited for parallel and distributed processing, and they naturally model multivariable systems. If a physical model is unavailable or too expensive to develop, a neural network model might be an alternative. There are now many real-world applications ranging from finance to aerospace. There are also many advanced neural network architectures (Haykin, 1994). Some representative neural network models will be discussed in the remainder of this chapter.

In terms of their architectures, neural networks can be categorized into feedforward and recurrent. The difference between these two is that there is at least one feedback loop in the latter. The focus of this book is on feedforward neural networks. Some representative feedforward models will be introduced in the following discussions.

1.2 Neural Network Learning

The primary significance of a neural network is its ability to learn from its environment. What is *learning*? In the context of neural networks, learning is defined as a process by which the free parameters of a neural network are adapted through a continuous process of stimulation by the environment. The type of learning is determined by the manner in which the parameter changes take place. The above definition implies that (1) the network is stimulated by the environment; (2) the network changes as a result of stimulation; and (3) the network responds to the environment in a new way after the occurrence of change.

1.2.1 Supervised Learning

During the training session of a neural network, an input is applied to the network, and a response of the network is obtained. The response is compared with an a priori target response. If the actual response differs from the target response, the neural network generates an error signal, which is then used to compute the adjustment that should be made to the network's synaptic weights so that the actual response matches the target output. In other words, the error is minimized, possibly to 0. Since the minimization process requires a teacher (supervisor), this kind of training is named *supervised learning*. A supervised learning model is illustrated in Fig. 1.2.

The notion of teacher comes from biological observations. For example, when learning a language, we hear the sound of a word from a teacher. The sound is stored in the memory banks of our brain, and we try to reproduce the sound. When we hear our own sound, we mentally compare it (actual response) with the stored sound (desired response) and note the error. If the error is large, we try again and again until it becomes significantly small. An unsupervised learning model is illustrated in Fig. 1.3.

1.2.2 Unsupervised Learning

In contrast to supervised learning, *unsupervised learning* does not require a teacher, i.e., there is no target response. During the training stage, the neural network receives its input patterns and it arbitrarily organizes them into categories. When an input is later applied, the neural network provides an output response to indicate the class to which the input pattern belongs. For example, show a person a set of different objects. Then ask him to separate them into different groups, such that objects in a group have one or more common features that distinguish them from other groups. When this (training) is done, show the same person an object that is unseen and ask him to place the object in one of the groups. He would then put it in the group with which the object shares the most common features.

Even though unsupervised learning does not require a teacher, it requires guidelines to determine how it forms groups. Grouping may be based on shape, color, or material consistency or some other properties of the objects. If no guidelines have

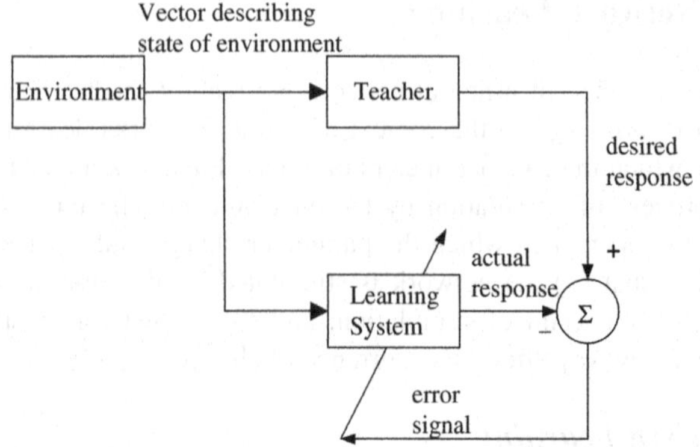

Fig. 1.2 Supervised learning model

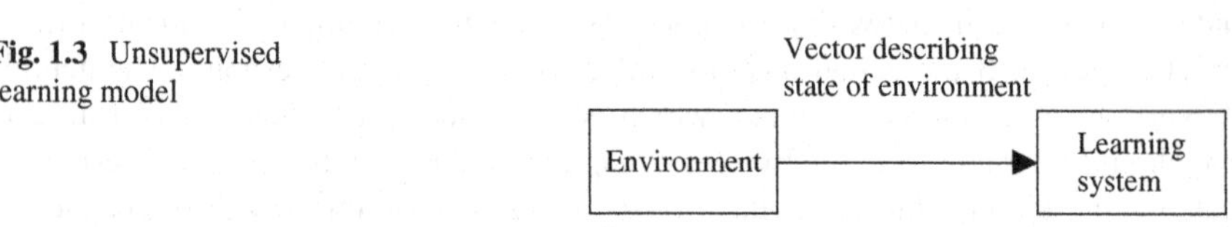

Fig. 1.3 Unsupervised learning model

been given as to what type of features should be used for grouping the objects, the grouping may or may not be successful. Similarly, to classify patterns effectively using neural networks, some guidelines are needed.

1.3 Perceptron

The Perceptron, the simplest form of a neural network, is able to classify data into two classes. Basically it consists of a single neuron with a number of adjustable weights.

The neuron is the fundamental processor of a neural network (Fig. 1.4). The summation in the neuron also includes an offset for lowering or raising the net input to the activation function. Mathematically the input to the neuron is represented by a vector $\mathbf{x} =< 1, x_1, x_2 \cdots x_n >^T$, and the output is a scalar y. The weights of the connections are represented by the vector $\mathbf{w} =< w_0, w_1, w_2 \cdots w_n >^T$, where w_o is the offset. The offset is often also called the *bias* and denoted with b.

The output is calculated as

$$y = f(\mathbf{w}^T\mathbf{x}). \tag{1.1}$$

Fig. 1.4 is a Perceptron with two inputs and an offset. With a hard limiter as activation function, the neuron produces an output equal to 0 or +1 that we can associate with the two classes C_1 and C_2 respectively.

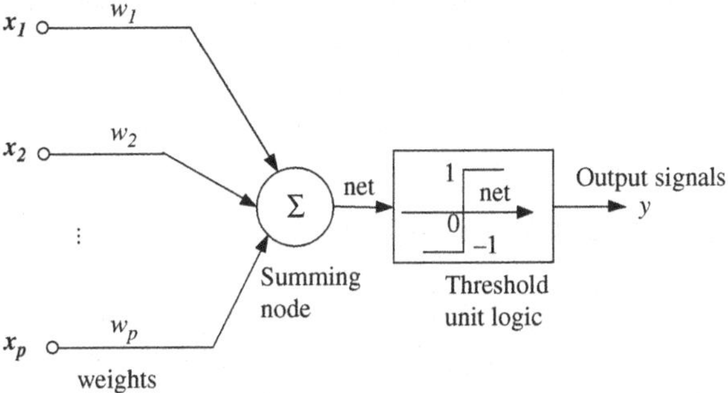

Fig. 1.4 Perceptron consisting of a neuron with an offset and a hard limiter activation function

The weights **w** are adjusted using an adaptive learning rule. One such learning rule is the *Perceptron Convergence Algorithm.* If the two classes C_1 and C_2 are linearly separable (i.e., they lie on opposite sides of a straight line or, in general, a hyperplane), then there exists a weight vector **w** with the properties

$$\begin{cases} \mathbf{w}^T\mathbf{x} \geq 0 & \text{for every input vector } \mathbf{x} \text{ belonging to class } C_1 \\ \mathbf{w}^T\mathbf{x} < 0 & \text{for every input vector } \mathbf{x} \text{ belonging to class } C_2 \end{cases} \tag{1.2}$$

Assuming, to be general, that the Perceptron has m inputs, the equation $\mathbf{w}^T\mathbf{x} = 0$ in an m-dimensional space with coordinates $x_1, x_2 \cdots x_m$, defines a hyperplane as the switching surface between the two different classes of input. The training process adjusts the weights **w** to satisfy the two inequalities (1.2). A training set consists of, say, K samples of the input vector **x** together with each sample's class membership (0 or 1). The learning is continued epoch by epoch until the weights stabilize. The core of the Perceptron convergence algorithm for adapting the weights of the elementary Perceptron has the following two steps.

Step 1. If the kth member of the training set, x_k ($k = 1, 2 \ldots$ K), is correctly classified by the weight vector $\mathbf{w}_k$ computed at the kth iteration of the algorithm, no correction is made to $\mathbf{w}_k$, i.e.,

$$\begin{cases} \mathbf{w}_{k+1} = \mathbf{w}_k & \text{if } \mathbf{w}^T\mathbf{x} \geq 0 \text{ and } \mathbf{x}_k \text{ belongs to class } C_1 \\ \mathbf{w}_{k+1} = \mathbf{w}_k & \text{if } \mathbf{w}^T\mathbf{x} < 0 \text{ and } \mathbf{x}_k \text{ belongs to class } C_2 \end{cases} \tag{1.3}$$

Step 2. Otherwise the Perceptron weights are updated according to the rule

$$\begin{cases} \mathbf{w}_{k+1} = \mathbf{w}_k - \eta_k\mathbf{x}_k & \text{if } \mathbf{w}^T\mathbf{x} \geq 0 \text{ but } \mathbf{x}_k \text{ belongs to class } C_2 \\ \mathbf{w}_{k+1} = \mathbf{w}_k + \eta_k\mathbf{x}_k & \text{if } \mathbf{w}^T\mathbf{x} < 0 \text{ but } \mathbf{x}_k \text{ belongs to class } C_1 \end{cases} \tag{1.4}$$

Notice the interchange of the class numbers from step 1 to step 2. The learning-rate η_k controls the adjustment applied to the weight vector at iteration k. If η_k is a constant, independent of the iteration number k, we have a fixed increment adaptation rule for the Perceptron. The algorithm has been proved to converge (Haykin, 1994).

1.4 Adaline and Least Mean Square Algorithm

ADALINE is an acronym for ADAptive LINear Element (also known as linear threshold unit). It was developed by Bernard Widrow and Marcian Hoff (1960). As shown in Fig. 1.5, the architecture of the Adaline is the same as that of the Perceptron. The difference between these two models is the activation function. An Adaline has n variable inputs $x_1, x_2 \cdots x_n$, which take on binary values of either +1 or -1. The bias input x_0 is fixed at a value of +1. Associated with the Adaline are n+1 adjustable analog weights $w_0, w_1, w_2 \cdots w_n$. The weights of the Adaline scale the corresponding inputs, the scaled inputs are summed, and the weighted sum is the input to a threshold device. The threshold device outputs a -1 for negative inputs and a +1 for positive inputs. The output of the threshold device is the Adaline output.

The Adaline applies the Least Mean Square (LMS) algorithm, also known as Widrow-Hoff learning. The LMS algorithm is more powerful than the Perceptron learning rule. While the Perceptron rule is guaranteed to converge to a solution that correctly categorizes the training patterns, the resulting network can be sensitive to noise, since patterns often lie close to the decision boundaries. The LMS algorithm minimizes mean square error, and therefore tries to move the decision boundaries as far from the training patterns as possible. The LMS algorithm has found many more practical uses than the Perceptron learning rule. This is especially true in the area of digital signal processing (Hagan, Demuth and Beale, 1996).

Given a training set $\{(\mathbf{x}_1, d_1), (\mathbf{x}_2, d_2) \ldots (\mathbf{x}_P, d_P)\}$, where the pth training pair is given by $(\mathbf{x}_p, d_p)$, with d_p being the target output, the mean square error (MSE) between the real output and the target output is calculated by:

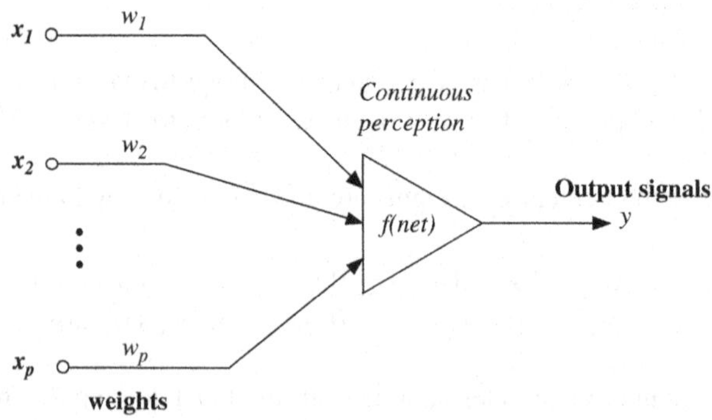

Fig. 1.5 Adaptive linear element (Adaline)

$$\begin{aligned} MSE &= E[e^2] = E[(d-y)^2] = E[(d-\mathbf{w}^T\mathbf{x})^2] \\ &= E\left[d^2 - 2d\mathbf{w}^T\mathbf{x} + \mathbf{w}^T\mathbf{x}\mathbf{x}^T\mathbf{w}\right] \\ &= E[d^2] - 2\mathbf{w}^T E[d\mathbf{x}] + \mathbf{w}^T E[\mathbf{x}\mathbf{x}^T]\mathbf{w} \end{aligned} \tag{1.5}$$

From Eq. (1.5), we can see that the MSE for the Adaline is a quadratic function:

$$J(\mathbf{w}) = c + \mathbf{b}^T\mathbf{w} + \frac{1}{2}\mathbf{w}^T\mathbf{A}\mathbf{w} \tag{1.6}$$

with $\mathbf{b} = -2\mathbf{h} = -2E\,[d\mathbf{x}]$ and $\mathbf{A} = 2\mathbf{R} = 2E\left[\mathbf{x}\mathbf{x}^T\right]$. The gradient of the MSE is given by:

$$\begin{aligned} \nabla J(\mathbf{w}) &= \nabla\left(c + \mathbf{b}^T\mathbf{w} + \tfrac{1}{2}\mathbf{w}^T\mathbf{A}\mathbf{w}\right) = \mathbf{b} + \mathbf{A}\mathbf{x} \\ &= -2\mathbf{h} + 2\mathbf{R}\mathbf{x} \end{aligned} \tag{1.7}$$

To find the optimal weights $\mathbf{w}^*$, we need to find the stationary point of $J(\mathbf{w})$. Let

$$\nabla J(\mathbf{w}) = -2\mathbf{h} + 2\mathbf{R}\mathbf{x} \triangleq 0 \tag{1.8}$$

So we have

$$\mathbf{w}^* = \mathbf{R}^{-1}\mathbf{h} \tag{1.9}$$

As the objective of neural network training is to find the optimal $\mathbf{w}^*$ by iterative updating, the steepest gradient method can be applied:

$$w(t+1) = w(t) - \eta\nabla J(t) \tag{1.10}$$

where t refers to the tth iteration. However, since it is not desirable to calculate $\mathbf{h}$ and $\mathbf{R}$, and not convenient to calculate $\mathbf{R}^{-1}$, the MSE can be estimated iteratively by square error (Widrow and Hoff, 1960):

$$\hat{J}(\mathbf{w}) = \frac{1}{2}[d_i(t) - y_i(t)]^2 \triangleq \frac{1}{2}e^2(t) \tag{1.11}$$

and its gradient can be accordingly estimated by

$$\begin{aligned} \nabla\hat{J}(\mathbf{w}) &= \frac{\partial\hat{J}(t)}{\partial\mathbf{w}(t)} = \frac{\frac{1}{2}\partial e^2(t)}{\partial\mathbf{w}(t)} = e(t)\cdot\frac{\partial e(t)}{\partial\mathbf{w}(t)} = e(t)\cdot\frac{\partial(d(t) - \mathbf{w}(t)\cdot\mathbf{x}(t))}{\partial\mathbf{w}(t)} \\ &= e(t)\cdot(-\mathbf{x}(t)) \end{aligned} \tag{1.12}$$

1.5 Multilayer Perceptron and Backpropagation Algorithm

As long as a neural network consists of a linear combiner followed by a nonlinear element, a single-layer Perceptron can perform pattern classification only on linearly separable patterns, regardless of the form of nonlinearity used. Linear separability requires that the patterns to be classified be sufficiently separated from each other to ensure that the decision boundaries are hyperplanes.

As shown in Fig. 1.6, MLP can address the above problem. An MLP consists of an input layer, one or more hidden layers and an output layer. Each neuron is fully connected to all the neurons in its preceding layer as well as all those in its next layer. Typically, the same nonlinear function, such as the log-sigmoid function, is applied to every neuron.

The backpropagation algorithm was developed for training multilayer Perceptron networks. It was popularized by Rumelhart, Hinton and Williams (1986), although similar ideas had been developed previously by others. The idea is to train a network by propagating the output errors backward through the layers. The errors serve to evaluate the derivatives of the error function with respect to the weights, which can then be adjusted.

The backpropagation algorithm consists of two phases, namely, a feedforward phase and a backpropagation phase. The former applies an input, evaluates the activations and stores the error. The latter computes the adjustments and updates the weights. However, only the errors in the output layer are visible, i.e., can be calculated directly, whereas those in the hidden layers will be propagated from their next layers.

The error at the output neuron j at iteration t can be calculated by the difference between the desired output and the corresponding real output, i.e., $e_j(t) = d_j(t) - y_j(t)$. Accordingly, the total error energy of all output neurons is $\varepsilon(t) = \frac{1}{2} \sum_{j \in C} e_j^2(t)$.

Referring to Fig. 1.6, the output of the kth neuron in the lth layer can be calculated by:

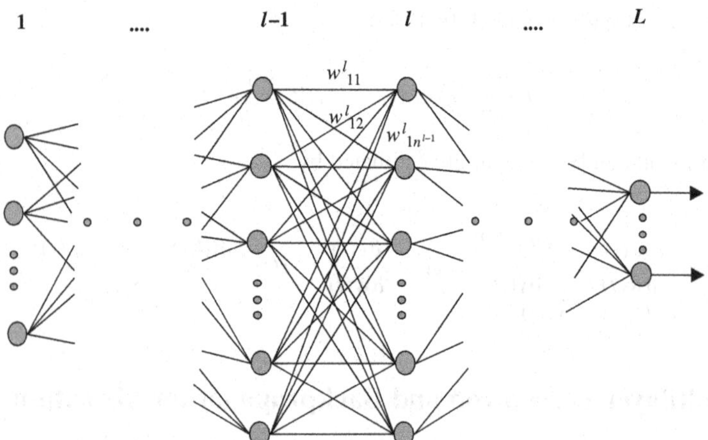

Fig. 1.6 Multilayer Perceptron

$$y_k^l = f\left(\sum_{j=1}^{n^{l-1}} w_{jk}^l \cdot y_j^{l-1}\right) \tag{1.13}$$

where $1 \le l \le L$, n^l refers to the number of neurons in layer l. For the input layer thus holds $l = 1$, $y_j^1 = x_j$; for the output layer $l = L$, $y_j^L = y_j$.

1.5.1 Output Layer Learning

The mean square error (MSE) of the output can be computed by:

$$E = \frac{1}{2}\sum_{j=1}^{n^L}(d_j - y_j)^2 = \frac{1}{2}\sum_{j=1}^{n^L}\left[d_j - f\left(\sum_{i=1}^{n^{L-1}} w_{ij}^L \cdot y_i^{L-1}\right)\right] \tag{1.14}$$

The steepest descent of MSE can be used to update the weights:

$$w_{ij}^L(t+1) = w_{ij}^L(t) - \eta\frac{\partial E}{\partial w_{ij}^L} \tag{1.15}$$

Since the output error E is an indirect function of the weights in the hidden layers, the chain rule of calculus is applied to calculate the derivatives. Let $\mathrm{net}_j^l = \sum_{i=1}^{n^{l-1}} w_{ij}^l \cdot y_i^{l-1}$, $y_j^l = f(\mathrm{net}_j^l)$; then we have

$$\frac{\partial E}{\partial w_{ij}^L} = \frac{\partial E}{\partial y_j^L}\cdot\frac{\partial y_j^L}{\partial w_{ij}^L}$$

$$[-12pt] = \frac{\partial E}{\partial y_j^L}\cdot\frac{\partial y_j^L}{\partial \mathrm{net}_j^L}\cdot\frac{\partial \mathrm{net}_j^L}{\partial w_{ij}^L} \tag{1.16}$$

$$[-12pt] = -(d_j - y_j)\cdot f'\left(\mathrm{net}_j^L\right)\cdot y_i^{L-1}$$

The error signal of the output can be backpropagated to layer $L-1$:

$$\delta_j^L = -\frac{\partial E}{\partial \mathrm{net}_j^L} = -\frac{\partial E}{\partial y_j^L}\cdot\frac{\partial y_j^L}{\partial \mathrm{net}_j^L} = (d_j - y_j)\cdot f'\left(\mathrm{net}_j^L\right) \tag{1.17}$$

1.5.2 Hidden Layer Learning

For any hidden layer l, the descent of the error is calculated by

$$\frac{\partial E}{\partial w_{ij}^{l}} = \frac{\partial E}{\partial y_j^{l}} \cdot \frac{\partial y_j^{l}}{\partial w_{ij}^{l}} = \frac{\partial E}{\partial y_j^{l}} \cdot \frac{\partial y_j^{l}}{\partial \mathrm{net}_j^{l}} \cdot \frac{\partial \mathrm{net}_j^{l}}{\partial w_{ij}^{l}} = \frac{\partial E}{\partial y_j^{l}} \cdot f'\left(\mathrm{net}_j^{l}\right) \cdot y_i^{l-1} \tag{1.18}$$

Since E is not a direct function of the output in the hidden layers, the steepest descent of the error can be calculated by:

$$\begin{aligned} \frac{\partial E}{\partial y_k^{l}} &= \sum_{k=1}^{n^{l+1}} \left(\frac{\partial E}{\partial y_k^{l+1}} \cdot \frac{\partial y_k^{l+1}}{\partial y_i^{l}} \right) \\ &= \sum_{k=1}^{n^{l+1}} \left(\frac{\partial E}{\partial y_k^{l+1}} \cdot \frac{\partial y_k^{l+1}}{\partial \mathbf{net}_k^{l+1}} \cdot \frac{\partial \mathbf{net}_k^{l+1}}{\partial y_i^{l}} \right) \\ &= \sum_{k=1}^{n^{l+1}} \left(e_k^{l+1} \cdot f'(\mathbf{net}_k^{l+1}) \cdot w_{jk}^{l+1} \right) = \sum_{k=1}^{n^{l+1}} \left(\delta_k^{l+1} \cdot w_{jk}^{l+1} \right) \end{aligned} \tag{1.19}$$

where δ_k^{l+1} is the propagated error signal from the next layer, and the error signal in the lth layer $\delta_k^{l} = e_k^{l} \cdot f'\left(\mathrm{net}_k^{l}\right)$ will then be propagated to layer $l-1$.

1.6 Radial Basis Function Networks

The design of a neural network can be viewed as a curve-fitting problem. According to this viewpoint, learning is equivalent to finding a surface in a multidimensional space that provides a best fit to the training data, whereas generalization is equivalent to the use of the trained multidimensional surface to interpolate the test data.

There are two bases in support of Radial Basis Function networks:

1. Cover's Theorem. *A complex pattern classification problem cast in a high-dimensional space nonlinearly is more likely to be linearly separable than in a low-dimensional space.* Cover's theorem tells us that we can map the input space to a high-dimensional space, in which a linear function will be found.
2. Tikhonov Regularization Theory. *In the context of a hypersurface reconstruction problem, the basic idea of regularization is to stabilize the solution by means of some auxiliary nonnegative functional that embeds prior information about the solution.* Tikhonov Regularization Theory tells us that we cannot purely rely on the training data, but must introduce some constraints, e.g., function smoothness.

RBF networks are feedforward networks trained using a supervised training algorithm. They are typically configured with a single hidden layer of units whose activation function is selected from a class of functions called basis functions. Actually, RBF networks transform the input space into a high dimensional space in a nonlinear manner and then find the curve-fitting approximation in high-dimensional space. The activation function depends on the Euclidean distance between input and target vectors. An RBF network can be described by

$$F(\mathbf{x}) = \sum_{i=1}^{n} w_i \varphi\left(\|\mathbf{x} - \mathbf{x}_i\|\right) \tag{1.20}$$

where φ is a nonlinear transformation from input space to hidden space of high dimensionality. The most common form of basis function is the Gaussian function

$$\varphi(\mathbf{x}) = \exp\left(-\frac{\|\mathbf{x} - c\|^2}{2\sigma^2}\right) \tag{1.21}$$

where c determines the center of the basis function, and σ is a width parameter that controls how spread out the curve is.

The output layer performs a linear mapping from hidden space to output space. The output of the network is the weighted sum of the hidden layer neuron outputs. A set of training data with their corresponding target outputs are given for supervised learning:

$$\begin{bmatrix} \varphi_{11} & \varphi_{12} & \cdots & \varphi_{1n} \\ \varphi_{21} & \varphi_{22} & \cdots & \varphi_{2n} \\ \vdots & \vdots & \vdots & \vdots \\ \varphi_{n1} & \varphi_{n2} & \cdots & \varphi_{nn} \end{bmatrix} \begin{bmatrix} w_1 \\ w_2 \\ \vdots \\ w_n \end{bmatrix} = \begin{bmatrix} d_1 \\ d_2 \\ \vdots \\ d_n \end{bmatrix} \tag{1.22}$$

So,

$$\mathbf{w} = \Phi^{-1}\mathbf{d} \tag{1.23}$$

In practice, we do not calculate Φ^{-1}, but apply the following two steps to train RBF networks: (1) Select centers randomly or based on some criteria, (2) update weights, typically using a linear least estimation algorithm.

1.7 Support Vector Machines

Given a number of n training samples for a two-class problem $\{(\mathbf{x}_1,y_1\}\cdots(\mathbf{x}_n,y_n\}$, where $\mathbf{x} \in R^m$, $y \in \{-1,+1\}$, there may exist a nested structure of decision functions

$$\{f_\alpha(x):\alpha \in \Lambda\} \text{ where } f_\alpha:R^m \to \{-1,+1\} \tag{1.24}$$

However, the optimal function should be the one that can achieve the maximum margin $\frac{2}{\|\mathbf{w}\|}$. The goal of support vector machines (SVMs) is to minimize $\|\mathbf{w}\|$ to get the maximum margin, as shown in Fig. 1.7.

The optimal solution f_α is supposed to provide the smallest possible value for expected risk $R(f_\alpha)$:

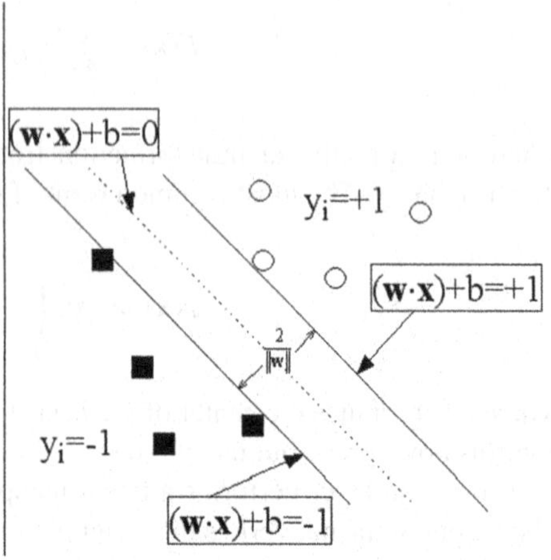

Fig. 1.7 Illustration of maximum margin

$$R(f_\alpha) = \int |f_\alpha(\mathbf{x}) - y| P(\mathbf{x},y) d\mathbf{x} dy \tag{1.25}$$

As the probability distribution $P(\mathbf{x},y)$ is unknown, the Empirical Risk, $\mathrm{R}_{\mathrm{emp}}(f_\alpha)$ is computed in all the conventional neural networks:

$$\mathrm{R}_{emp}(f_\alpha) = \frac{1}{n}\sum_{i=1}^{n}\frac{1}{2}|f_\alpha(\mathbf{x}_i - y_i)| \tag{1.26}$$

However, in terms of Vapnik-Chervonenkis (VC) Dimension theory (Vapnik, 1995), which indicates the maximum number of training points h that can be separated by set of linear functions f_α, there is a bound to the expected risk:

$$R(f_\alpha) \leq \mathrm{R}_{\mathrm{emp}}(f_\alpha) + \sqrt{\frac{h\left(\ln\frac{2\mathrm{n}}{\mathrm{h}} + 1\right) - \ln\frac{\eta}{4}}{n}} \tag{1.27}$$

with probability $1 - \eta$. The second term in Eq. (1.27) is called the confidence term or confidence interval. For a fixed number of training examples n, the training error (empirical risk) decreases monotonically as the capacity or VC dimension h is increased, whereas the confidence interval increases monotonically. Accordingly, both the guaranteed risk and the generalization error go through a minimum. Hence, our goal is to find a network such that decreasing the VC dimension occurs at the expense of the smallest possible increase in training error.

So the goal of SVMs is to minimize $\|\mathbf{w}\|^2$ subject to $y_i(\mathbf{w}\cdot\mathbf{x}_i + b) \geq 1$, $i = 1,\ldots,n$. This is equivalent to solving the following constrained Quadratic Programming (QP) problem to construct the Lagrangian:

$$L(\mathbf{w},b,\Lambda) = \frac{1}{2}\|\mathbf{w}\|^2 - \sum_{i=1}^{n} \alpha_i \left[y_i (\mathbf{w} \cdot \mathbf{x}_i + b) - 1 \right] \tag{1.28}$$

where $\Lambda = (\alpha_1, ..., \alpha_n)$ and $\Lambda \geq 0$.

A solution to the QP problem is determined by the saddle point of the function. Differentiate $L(\mathbf{w},b,\Lambda)$with respect to $\mathbf{W}$ and b.

$$\frac{\partial L(\mathbf{w},b,\Lambda)}{\partial \mathbf{w}} = \mathbf{w} - \sum_{i=1}^{n} \alpha_i y_i x_i = 0 \Rightarrow \mathbf{w} = \sum_{i=1}^{n} \alpha_i y_i x_i \tag{1.29}$$

$$\frac{\partial L(\mathbf{w},b,\Lambda)}{\partial b} = \sum_{i=1}^{n} \alpha_i y_i = 0 \tag{1.30}$$

So, the optimization problem is converted to maximizing

$$F(\Lambda) = \sum_{i=1}^{n} \alpha_i - \frac{1}{2} \sum_{i,j=1}^{n} \alpha_i \alpha_j y_i y_j \mathbf{x}_i \cdot \mathbf{x}_j \tag{1.31}$$

subject to $\Lambda \cdot \mathbf{y} = 0$ where $\Lambda \geq 0$. In most cases $\alpha = 0$. Those input data points with $\alpha > 0$ are called support vectors. Removing other data points except support vectors will not affect the solution of the classifier.

Similarly to RBF, the input space is also mapped to a sufficiently large feature space in SVM, so that patterns become linearly separable, and so a simple Perceptron in feature space can do the classification. SVMs are radically different types of classifiers which have attracted a great deal of attention lately due to the novelty of the concepts that they bring to pattern recognition, their strong mathematical foundation, and their excellent results in practical problems.

Chapter 2
Principles of Sensitivity Analysis

Sensitivity refers to how a neural network output is influenced by its input and/or weight perturbations. Sensitivity analysis dates back to the 1960 s, when Widrow investigated the probability of misclassification caused by weight perturbations, which are caused by machine imprecision and noisy input (Widrow and Hoff, 1960). In network hardware realization, such perturbations must be analyzed prior to its design, since they significantly affect network training and generalization. The initial idea of sensitivity analysis has been extended to the optimization of neural networks, such as through sample reduction, feature selection, and critical vector learning.

2.1 Perturbations in Neural Networks

Perturbations of neural networks are caused by machine imprecision and/or input noise. For the purpose of analysis, these perturbations can be simulated by embedding disturbance to the original inputs or connection weights. Perturbation analysis allows the study of the characteristics of a function under small perturbations of the function's parameter (Holtzman, 1992; Zurada et al., 1997). Perturbation analysis is also important also when assessing network robustness against input noise to measure the uncertainty or fluctuations. In perturbation analysis we are interested in evaluating the disturbance in the function's response to small perturbations in its parameters. Fig. 2.1 shows how to investigate the effects of perturbations to neural networks.

Assuming that the performance function is differentiable, the relationship between the perturbed response of this function and parameter perturbations is expressed by a Taylor expansion of the function. For example, for a one-dimensional cost function g

$$g(\Theta + \Delta\Theta) = g(\Theta) + \frac{\Delta\Theta}{1!} g'(\Theta) + \frac{\Delta\Theta^2}{2!} g''(\Theta) + \cdots \tag{2.1}$$

where Θ is a parameter of the function and typically includes weight $\mathbf{W}$ and input $\mathbf{X}$; $\Delta\Theta$ is a small perturbation of Θ. The performance cost function g is usually taken

D.S. Yeung et al., *Sensitivity Analysis for Neural Networks*, Natural Computing Series, DOI 10.1007/978-3-642-02532-7_2, © Springer-Verlag Berlin Heidelberg 2010

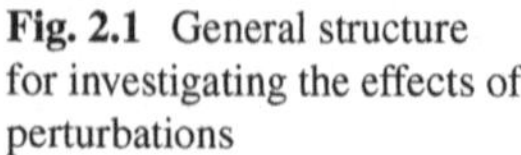

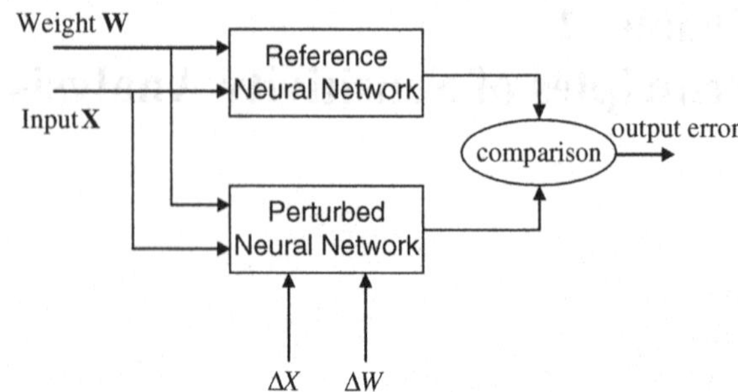

Fig. 2.1 General structure for investigating the effects of perturbations

by the noise-to-signal ratio (NSR) or expectation of decision errors. The Taylor expansion shows that the derivatives of the function with respect to the perturbed parameter encapsulate the characteristics of that function under the perturbations. Ideally, when $\Delta\Theta \to 0$, $\frac{\Delta\Theta}{1!}g'(\Theta) + \frac{\Delta\Theta^2}{2!}g''(\Theta) + \cdots \to 0$.

Eq. (2.1) shows that the derivatives play a very important role in determining the influence of parameter perturbations on the output of the performance function. Sensitivity analysis is applied to investigate how the derivatives can be used to quantify the response of the system to parameter perturbations, and how these derivatives can be calculated.

Sensitivity analysis techniques differ mainly in the cost function used, the order of the derivatives that are considered, whether the analysis is in continuous time or for discrete time intervals, and the way in which the derivatives are calculated. Due to computational considerations, sensitivity analysis is based on approximations of Eq. (2.1), usually first-order or second-order approximations. The higher the order of the approximation, the more accurate but more complex and time consuming the process. Usually, sensitivity analysis is done at discrete time intervals for that time interval only. Sensitivity analysis can also be performed for continuous time models, referred to as stochastic analysis (Koda, 1995; 1997).

2.2 Neural Network Sensitivity Analysis

Without loss of generality, let us consider a neural network performing a nonlinear, differentiable mapping $\Gamma:\Re^I \to \Re^K$, from input $\mathbf{x} = (x_1, x_2 \ldots x_I)$ to output $o = (o_1, o_2 \ldots o_K)$. Suppose $\mathbf{x}^{(n)} \in \Omega$, where Ω is an open set. Since o is differentiable at $\mathbf{x}^{(n)}$ we have

$$o(x + \Delta x) = o(x^{(n)}) + J(x^{(n)})\Delta x + g(\Delta x) \tag{2.2}$$

where

$$J(x^{(n)}) = \begin{bmatrix} \frac{\partial o_1}{\partial x_1} & \frac{\partial o_1}{\partial x_2} & \cdots & \frac{\partial o_1}{\partial x_I} \\ \frac{\partial o_2}{\partial x_1} & \frac{\partial o_2}{\partial x_2} & \cdots & \frac{\partial o_2}{\partial x_I} \\ \vdots & \vdots & \vdots & \vdots \\ \frac{\partial o_K}{\partial x_1} & \frac{\partial o_K}{\partial x_2} & \cdots & \frac{\partial o_K}{\partial x_I} \end{bmatrix} \tag{2.3}$$

is the Jacobian matrix and

$$\lim_{\Delta x \to 0} \frac{g(\Delta x)}{|\Delta x|} = 0 \tag{2.4}$$

Fig. 2.2 illustrates geometrical interpretation of Eq. (2.2) in space $\Re^K$. Point $o(x^{(n)})$ represents the nominal response of the neural network for the nth element of the training set $\mathbf{x}^{(n)}$. The disturbance Δx of the input vector causes the perturbed response at $o(x(n) + \Delta x)$. This response can be expressed as a combination of three vectors as indicated in Eq. (2.2).

The sensitivity analysis can be used for different purposes in neural networks (Engelbrecht, 1999):

Optimization. The calculation of the gradient of a function forms an important part of optimization. One of the first uses of sensitivity analysis is therefore in optimization problems (Cao, 1985; Holtzman, 1992). In neural networks, derivatives of the objective function with respect to the weights are computed to locate minima by driving these derivatives to 0. Second order derivatives have also been used to develop more sophisticated optimization techniques to improve convergence and accuracy. Koda (1995, 1997) employed stochastic sensitivity analysis to compute the gradient for time-dependent networks such as recurrent neural networks.

Robustness. Neural network robustness and stability analysis is the study of the conditions under which the outcome of the neural network changes. This study is important for hardware implementation of neural networks to ensure stable

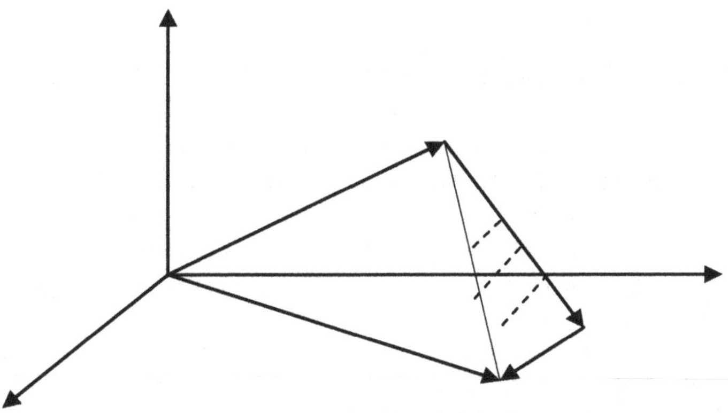

Fig. 2.2 Illustration of output disturbance caused by input disturbance

networks that are not adversely affected by weight, external input and activation function perturbations (Alippi, Piuri and Sami, 1995; Oh and Lee, 1995). Instead of using derivatives to compute the gradient of the objective function with respect to the weights, Jabri and Flower (1991) use differences to approximate the gradient, thereby significantly reducing hardware complexity.

Generalization. Fu and Chen (1993) state good generalization must imply insensitivity to small perturbations in inputs. They derive equations to compute the sensitivity of the neural network output vector to changes in input values, and show under what conditions global neural network sensitivity can be reduced. For example, using small slopes for the sigmoid activation function, using as small as possible weights, reducing the number of units, and ensuring activation levels close to 0 or 1 will reduce network sensitivity. Choi and Choi (1992) derive a neural network sensitivity norm which expresses the sensitivity of the neural network output with respect to input perturbations. This neural network norm is then used to select from sets of optimal weights the weight set with lowest neural network sensitivity, which results in the best generalization.

Measure of nonlinearity. Lamers, Kok and Lebret (1998) use the variance of the sensitivity of the neural network output to input parameter perturbations as a measure of the nonlinearity of the data set. This measure of nonlinearity is then used to show that the higher the variance of noise injected to output values, the more linearized the problem.

Causal inference. Sensitivity analysis has been used to assess the significance of model inputs. Engelbrecht, Cloete and Zurada (1995) use exact derivative calculations to compute the significance values which have a high influence on the neural network output. Goh (1993) derived a similar method using differences to approximate the gradient of the neural network output function with respect to inputs.

Selective learning. Hunt and Deller (1995) use weight perturbation analysis to determine the inference each pattern has on weight changes during training. Only patterns that exhibit a high influence on weight changes are used for training. Engelbrecht (1998) presents new active learning models based on sensitivity analysis, which use a measure of pattern informativeness to dynamically select patterns during training.

Decision boundary visualization. Goh (1993) uses an approximation to the derivatives of the neural network output function with respect to inputs to graphically visualize decision boundaries. Engelbrecht (1998) shows how exact derivative calculations can be used to locate and visualize decision boundaries. Victor (1998) uses the decision boundary algorithm to improve the accuracy of rules extracted from trained neural networks in a cooperative learning environment.

Pruning. Sensitivity analysis has been applied extensively to neural network pruning. One technique is to compute the sensitivity of the objective function with respect to neural network parameters (Le Cun, 1990; Moody et al., 1995). Another method of sensitivity analysis pruning is to compute the sensitivity of the neural network output function to parameter perturbations (Zurada, 1994, 1997).

Learning derivatives. Basson and Engelbrecht (1999) developed a new learning algorithm for feedforward neural networks that also learns the first-order derivatives of the neural network output with respect to each input unit while learning the underlying function. The neural network consists of two parts, one representing the learned function, and the other representing the derivatives of the learned function. Concepts from sensitivity theory are used to create a training set for the training of the derivative part of the neural network using gradient descent.

2.3 Fundamental Methods of Sensitivity Analysis

In system sensitivity theory, some sensitivity functions are introduced, such as the output sensitivity, the trajectory sensitivity and the performance-index sensitivity functions (Frank, 1978). All the methods of neural network sensitivity analysis can be divided into two categories, namely, geometrical approach and statistical approach. In 1962, Hoff used an n-dimensional hypersphere to model Adaline for sensitivity analysis (Hoff, 1962), which was further simplified in Glanz (1965). After two decades, Winter (1989) was the first one to derive an analytical expression for the probability of error in Madaline caused by weight perturbations. Stevenson continued Winter's work and established the sensitivity of Madaline to weight error (Stevenson, 1990; Stevenson, Winter and Widrow, 1990). A milestone work was done by Piché, who used a statistical approach to relate the output error to the change of weights for an ensemble of Madalines, with several activation functions such as linear, sigmoid, and threshold (Piché, 1992; Piché, 1995).

Figure 1.6 can be treated as a general neural network model. A neural network can have L layers, and each layer l $(0 \leq l \leq L)$ has $n^l (n^l \geq 1)$ neurons. n^0 stands for the input layer and n^L for the output layer. Since the number of neurons in layer $l-1$ is equal to the output dimension of that layer, which is also equal to the input dimension of layer l, the input dimension of layer l is n^{l-1}. For a neuron i $(1 \leq i \leq n^l)$ in layer l, its input vector, weight vector and output are $X^l = \left(x_1^l \cdots x_{n^{l-1}}^l\right)^T$, $W_i^l = \left(w_{i1}^l \cdots w_{in^{l-1}}^l\right)^T$ and $y_i^l = f(X^l \cdot W^l)$ respectively, where $f(\cdot)$ is an activation function. For each layer l, its input vector is X^l, its weight set is $W^l = \{W_1^l \cdots W_{n^l}^l\}$, and its output vector is $Y^l = \left(y_1^l \cdots y_{n^l}^l\right)^T$. For the network, its input is the vector X^1 or Y^0, its weight is W, and its output is Y^L. Let $\Delta X^l = \left(\Delta x_1^l \cdots \Delta x_{n^{l-1}}^l\right)^T$ and $\Delta Y^l = \left(\Delta y_1^l \cdots \Delta y_{n^l}^l\right)^T$ be the corresponding deviations for input and output, respectively.

2.3.1 Geometrical Approach

The use of n-dimensional geometry has proven to be a valuable tool for understanding and analyzing the Adaline. The geometrical interpretation of the equation dictating the Adaline's input-output map is the bias for most of the analysis presented in Stevenson (1990).

The input vector X in a neural network can be treated as a vector from the origin to the point $(x_1, x_2 \ldots x_n)$ in n-space. The point $(x_1, x_2 \ldots x_n)$ will be referred to "the tip of X" in the following discussion. In n-space, points at a distance r from the point c form a hypersphere of radius r centered at c. The surface area of such a hypersphere $A_n(r)$ is

$$A_n(r) = 2\sqrt{\pi^n}\Gamma\left(\frac{n}{2}\right) \cdot r^{n-1} \tag{2.5}$$

where $\Gamma(\bullet)$ is the Gamma function. As shown in Fig. 2.3, the connection weights, which are a vector at angle θ to the input vector $X = (x_1, x_2 \ldots x_n)$ in n-space, satisfy

$$X \cdot W \triangleq \sum_{i=1}^{n} x_i w_i = c \tag{2.6}$$

which is called a hyperplane for some scalar c. This hyperplane is perpendicular to the vector W and is at a distance $c/|W|$ from the origin. Particularly, HP_w is one of the hyperplanes passing through the origin with $c = 0$. Similarly to HP_w, HP_X denotes a hyperplane passing through the origin and perpendicular to the vector X. Hence, both HP_w and HP_X divide the hypersphere in two hemi hyperspheres, i.e., $H_w{}^+$ versus $H_w{}^-$, and $H_X{}^+$ versus $H_X{}^-$, respectively. There are four lunes intersected by these four hemi hyperspheres, $H_X^+ \cap H_W^+$, $H_X^+ \cap H_W^-$, $H_X^- \cap H_W^+$ and $H_X^- \cap H_W^-$, as shown in Fig. 2.3.

As the angle between X and W is θ, both the intersections, $H_X^+ \cap H_W^-$ and $H_X^- \cap H_W^+$, describe lunes of angle θ whereas the intersections $H_X^+ \cap H_W^+$ and $H_X^- \cap H_W^-$ both describe lunes of angle $(\pi - \theta)$. The ratio of the surface content of a lune of angle θ to the surface content of the entire hypersphere is $\theta/2\pi$.

Assuming binary-valued inputs, there are 2^n possible input patterns for an Adaline with n variable inputs. Each input pattern corresponds to a point in n-space which lies on a hypersphere of radius $\sqrt{n}$ centered at the origin. The Hoff

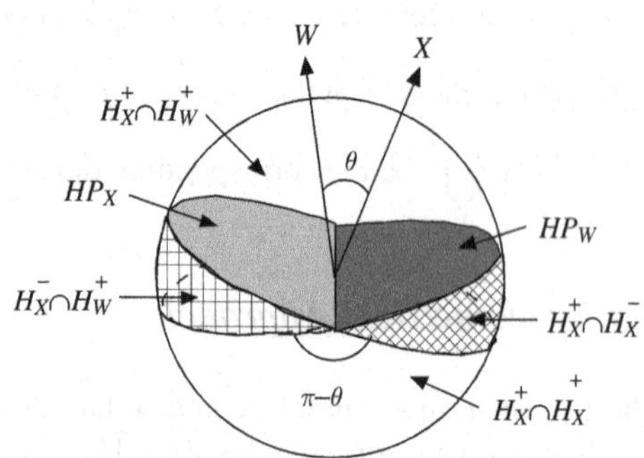

Fig. 2.3 Hypersphere approximation for sensitivity analysis

hypersphere-area approximation states that as n gets large, the points corresponding to the n-dimensional input patterns are approximately uniformly distributed over the surface of a hypersphere in n-space. Consequently, the percentage of input patterns which correspond to points on a selected region of the hypersphere can be approximated as the ratio of the surface content of the selected region to the surface content of the entire hypersphere (Hoff, 1962).

2.3.2 Statistical Approach

Let us consider the behavior of a multi-input/single-output mapping first. Let us assume that the connection weight vector changes from W^* to $W = W^* + \Delta W$, where ΔW indicates weight perturbations. Under the assumption of statistical weight perturbations, the statistical sensitivity can be defined as follows (Choi and Choi, 1992):

$$S^p(W^*) = \lim_{\sigma \to 0} \frac{\sqrt{\mathrm{var}[\Delta X_p(L)]}}{\sigma} \tag{2.7}$$

where $\Delta X_p(L)$ is the output error, σ is the standard deviation of each component of deviation of each component of ΔW, and var[...] is the variance. The output error vector $\Delta X_p(L)$ in the lth layer arising from weight perturbations ΔW is given by

$$\Delta X_p(l) = X_p(l) - X_p^*(l) \cong \sum_{k=1}^{l} C_{k,l}\left(W^*, X_p^*\right) \Delta W(k) X_p^*(k-1) \tag{2.8}$$

where the $N_l \times N_k$ matrix $C_{k,l}\left(W^*, X_p^*\right)$ is defined as

$$C_{k,l}\left(W^*, X_p^*\right) = \prod_{n=k+1}^{l} \nabla T_n W_n^* \nabla T_k = \nabla T_l W_l^* \nabla T_{l-1} W_{l-1}^* \ldots \nabla T_{k+1} W_{k+1}^* \tag{2.9}$$

Here, $\mathbf{T}_\mathrm{k}$ refers to the nonlinear transformation associated with the kth layer, and ∇T_k refers to its derivation with respect to the output of the kth layer.

As in the case of the weight perturbation discussed above, the sensitivity to input perturbation can be also calculated by Eq. (2.7) with σ being the standard deviation of each component of the input perturbation $\Delta X_p(0)$. In this respect, the output error $\Delta X_p(L)$ in output layer L caused by input perturbation is given by:

$$\Delta X_p(L) = C_{L,0}\left(W^*, X_p^*\right) \Delta X_p(0) \tag{2.10}$$

To measure sensitivity for a multi-output neural network, typically, the sensitivity of each output neuron is calculated first, and then the final sensitivity is set as the maximum or the average value among all output neurons.

Piché (1995) introduced a stochastic model for the statistical analysis of sensitivity. He assumed that (1) over the ensemble of networks, the weights of a particular layer all have the same variance, (2) the weights in the networks are statistically independent and (3) the mean value of each weight of the network over the ensemble is 0. Under these assumptions, it is possible to model the ensemble of weights associated with each node of the network as a vector of random variables. He discussed the selection of weight accuracies for Madaline using a statistical approach to sensitivity analysis. According to his stochastic model, all neurons have the same activation function. All network inputs, weights, input perturbations, and weight perturbations are random variables. The sensitivity of Madaline in this model is defined as the NSR of the output layer; the output NSR of a layer of threshold Adaline is:

$$NSR = \frac{\sigma_{\Delta y}^2}{\sigma_y^2} = \frac{4}{\pi}\sqrt{\frac{\sigma_{\Delta x}^2}{\sigma_x^2} + \frac{\sigma_{\Delta w}^2}{\sigma_w^2}} \tag{2.11}$$

where σ_y^2, σ_x^2, σ_w^2, $\sigma_{\Delta y}^2$, $\sigma_{\Delta x}^2$ and $\sigma_{\Delta w}^2$ refer to the variances of output y, input x, weight w, output error Δy, input perturbation Δx and weight perturbation Δw, respectively.

2.4 Summary

Sensitivity is initially investigated for the realization of a neural network to calculate its output perturbation caused by machine imprecision and noisy input. It is particularly useful and valuable to apply sensitivity analysis to the software design of networks, in which artificial perturbation is embedded in the training. In this paper, the principle of sensitivity analysis is discussed in detail, followed by systematic introduction to the advanced research in this area over the last two decades. All the existing techniques can be categorized into geometrical and statistical approaches. The former use hypersphere or hyper-rectangle model in n-dimensional space to analyze the deviation of the output vector caused by the perturbation of the input and weights. The latter calculate the sensitivity by noise-to-signal ratio or expectation of the output deviation.

Chapter 3
Hyper-Rectangle Model

In this chapter, we discuss a hyper rectangle model, instead of the traditional hypersphere, which is employed as the mathematical model to represent an MLP's input space. The hyper-rectangle approach does not demand that the input deviation be very small as the derivative approach requires, and the mathematical expectation used in the hyper-rectangle model reflects the network's output deviation more directly and exactly than the variance does. Moreover, this approach is applicable to the MLP that deals with infinite input patterns, which is an advantage of the MLP over other discrete feedforward networks like Madalines.

3.1 Hyper-Rectangle Model for Input Space of MLP

Neurons in an MLP, like cells, are organized into layers by linking them together. Links exist only between neurons of two adjacent layers; there is no link between neurons in the same layer or in any two nonadjacent layers. All neurons in a layer are fully linked to all the neurons in the immediately preceding layer and to all the neurons in the immediately succeeding layer. At each layer except the input layer, the inputs of each neuron are the outputs of the neurons in the previous layer, $X^l = Y^{.l-1} (l \geq 1)$.

Through training, a hyperplane P is obtained for classification or regression

$$P\colon \sum_{j=1}^{n} x_j w_j + \sigma = 0 \quad \text{in input space } \Omega \tag{3.1}$$

The computation of the sensitivity is essentially reduced to the calculation of the following integral (Zeng and Yeung, 2003):

$$I\,(A,B,W,\sigma) = \int_{a_1}^{b_1} \int_{a_2}^{b_2} \cdots \int_{a_n}^{b_n} \frac{\varphi(X)}{1 + \exp\left(-\sum\limits_{j=1}^{n} x_j w_j - \sigma\right)} dx_1 dx_2 \cdots dx_n. \tag{3.2}$$

D.S. Yeung et al., *Sensitivity Analysis for Neural Networks*, Natural Computing Series, DOI 10.1007/978-3-642-02532-7_3, © Springer-Verlag Berlin Heidelberg 2010

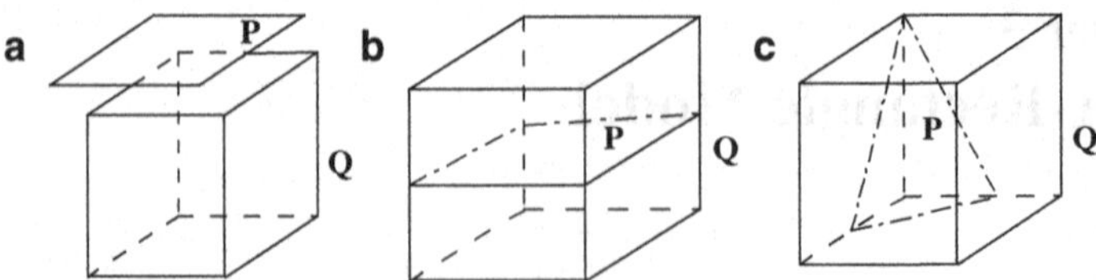

Fig. 3.1 Spatial relationship between P and Ω. (**a**) P does not cut Ω. (**b**) P cuts the four parallel sides of Ω. (**c**) P cuts a prism from Ω

where $A = \{a_1, a_2 \dots a_n\}$ and $B = \{b_1, b_2 \dots b_n\}$ are sets of lower and upper bounds of the integral, and $W = \{w_1, w_2 \dots w_n\}$ is the set of the weights, with $b_j > a_j$ and $w_j = 0$ for $1 \le j \le n$. It can be clearly seen that the input space Ω can be regarded as a hyper-rectangle bounded by A and B in n-dimensional space

From a geometric point of view, the hyperplane P can be regarded as a dividing plane. It divides the n-dimensional space Ω into three parts: the points on P, the points below P, and the points above P. The possible spatial relationship between P and Ω is illustrated in Fig. 3.1.

3.2 Sensitivity Measure of MLP

Based on the above hype-rectangle model, the sensitivity measure of Eq. (3.2) can be rewritten as:

$$I \approx \begin{cases} \int\limits_{\Omega} \cdots \int \left(\sum\limits_{k=0}^{p} (-1)^k \exp\left(-k \sum\limits_{j=1}^{n} x_j w_j - k\sigma \right) \right) d\Omega & \text{if} \sum\limits_{j=1}^{n} x_j w_j + \sigma > 0 \\ \int\limits_{\Omega} \cdots \int i^{p_1}(X) d\Omega + \int\limits_{\Omega} \cdots \int i^{p_2}(X) d\Omega & \text{if} \sum\limits_{j=1}^{n} x_j w_j + \sigma = 0 \\ \int\limits_{\Omega} \cdots \int \left(\sum\limits_{k=0}^{p} (-1)^k \exp\left(-k \sum\limits_{j=1}^{n} x_j w_j + k\sigma \right) \right) d\Omega & \text{if} \sum\limits_{j=1}^{n} x_j w_j + \sigma < 0 \end{cases} \tag{3.3}$$

where $i^p(X)$ denotes the first p-term Taylor expansion of the integrand. A solution to this problem is to locate the intersection points of P and those parallel lines that are the extensions of the edges of Ω. Each line must have one and only one intersection point with P under the assumption $w_j \neq 0$ for $1 \le j \le n$. By comparing the coordinates on a given axis of those intersection points with the lower bound and the upper bound of Ω on that axis, it is easy to identify the intersection situations between P and Ω. It is known that Ω has 2^{n-1} edges parallel to a given coordinate axis. Take the x_n-axis as an example. Each edge parallel to the x_n-axis can be determined by assigning a total of $n-1$ coordinates $(x_1 \dots x_{n-1})$ with either $x_j = a_j$ or $x_j = b_j$. Thus, the x_n-coordinate for a given line's intersection point can be calculated by

$$x_n = \left(\sum_{j=1}^{n-1} x_j w_j + \sigma \right) \Big/ w_n \tag{3.4}$$

with its $n-1$ fixed coordinates, which consequently yield a set $E_n = \{e_{n,1}, e_{n,2} \ldots e_{n,2^{n-1}}\}$ representing 2^{n-1} edges.

3.3 Discussion

The sensitivity measure is something like an analog rule that network users can use as a tool to evaluate their network's performance. For example, Lee and Oh (1994) indicated that in pattern classification applications, their sensitivity (misclassification probability) results could be applied to select the optimal weight set among many weight sets acquired through a number of learning trials. Among these weight sets, an MLP with the optimal set will have the best generalization capability and the best input noise immunity. Zurada, Malinowski, and Usui (1997) made use of analytical sensitivity to delete redundant inputs to an MLP for pruning its architecture. Engelbrecht and Cloete (1999) also used analytical sensitivity to dynamically select training patterns.

In this chapter, sensitivity is analyzed through the hyper-rectangle model, which is especially suitable for the most popular and general feedforward network: multilayer Perceptron. The sensitivity measure is defined as the mathematical expectation of output deviation due to expected input deviation with respect to overall input patterns in a continuous interval. Based on the structural characteristics of the MLP, a bottom-up approach is adopted. A single neuron is considered first, and algorithms with approximately derived analytical expressions that are functions of expected input deviation are given for the computation of its sensitivity. Then another algorithm is given to compute the sensitivity of the entire MLP network.

Chapter 4
Sensitivity Analysis with Parameterized Activation Function

Among all the traditional methods introduced in Chap. 2, none has involved activation function in the calculation of sensitivity analysis. This chapter attempts to generalize Piché's method by parameterizing antisymmetric squashing activation functions, through which a universal expression of MLP's sensitivity will be derived without any restriction on input or output perturbations.

4.1 Parameterized Antisymmetric Squashing Function

Any antisymmetric squashing activation function $g(x)$ can be specified by three parameters, and the main idea of the function approximation approach is to use another function $g_{A,B,C}(x)$ to approximate $g(x)$. The function has the following form:

$$g_{A,B,C}(x) = g^{+}_{A,B,C}(x) + g^{-}_{A,B,C}(x) \tag{4.1}$$

where

$$g^{+}_{A,B,C}(x) = \begin{cases} B - \dfrac{B-C}{2}e^{-Ax^2}, & \text{if } x \geq 0 \\ 0, & \text{if } x < 0 \end{cases} \tag{4.2}$$

$$g^{-}_{A,B,C}(x) = \begin{cases} 0, & \text{if } x \geq 0 \\ C + \dfrac{B-C}{2}e^{-Ax^2}, & \text{if } x < 0 \end{cases} \tag{4.3}$$

The objective of function approximation is to determine the three factors A, B and C of $g(x)$, so that the distance between the two functions $g_{A,B,C}(x)$ and $g(x)$ is minimized. Here the distance is defined in terms of the L^2 norm. The factors A, B and C are determined by

$$\begin{cases} \min\limits_{A} \sqrt{\int_0^{+\infty} \left(g(x) - g_{A,B,C}(x)\right)^2 dx} \\ B = \lim\limits_{x \to +\infty} g(x) \\ C = -B \end{cases} \tag{4.4}$$

D.S. Yeung et al., *Sensitivity Analysis for Neural Networks*, Natural Computing Series, DOI 10.1007/978-3-642-02532-7_4, © Springer-Verlag Berlin Heidelberg 2010

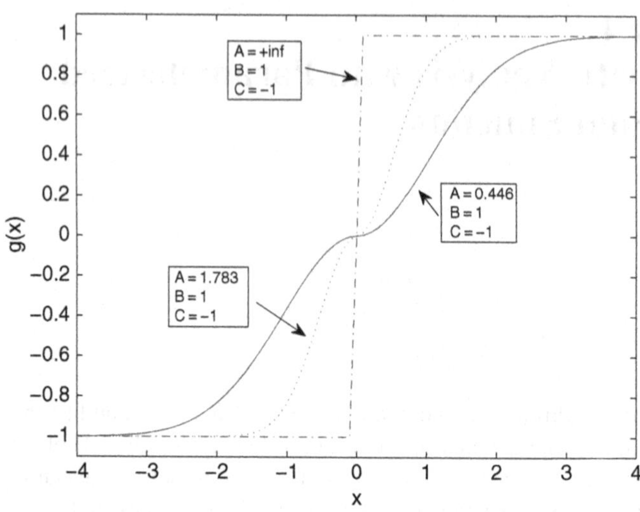

Fig. 4.1 Approximation results of some antisymmetric squashing activation functions

Because both $g_{A,B,C}(x)$ and $g(x)$ are antisymmetric about the point (0, 0), the factor A can be found by solving the minimization problem over $(0, +\infty)$ instead of $(-\infty, +\infty,)$. Some commonly used antisymmetric squashing functions are $g(x) = \frac{e^x - e^{-x}}{e^x + e^{-x}}$ or $g(x) = \frac{e^x - 1}{e^x + 1}$. The approximated results are shown in Fig. 4.1.

From Eqs. (4.1), (4.4) and Fig. 4.1, the following can be observed. (1) When a parameter changes from 0 to $+\infty$, the function $g_{A,B,C}(x)$ becomes the threshold activation function, and we need not solve the threshold activation function and the sigmoidal function separately. (2) This function approximation approach still cannot avoid inaccuracy, but it helps us obtain an analytical expression of an MLP's sensitivity for a class of activation functions. Then the effect of the activation function on the sensitivity of an MLP can be investigated. The function approximation approach is a tradeoff between accuracy and generalization.

4.2 Sensitivity Measure

According to Piché (1995), the sensitivity of the jth neuron in the lth layer is calculated by the NSR of the output

$$S_R(y_j^l) = \sqrt{\frac{D(\Delta y_j^l)}{D(y_j^l)}} \tag{4.5}$$

where

$$D(y_j^l) = \int_{-\infty}^{+\infty} g^{2(\alpha u)} \cdot \frac{1}{\sqrt{2\pi}} e^{-(u^2/2)} du - E^2(y_j^l)$$
$$D(\Delta y_j^l) = \int_{-\infty}^{+\infty}\int_{-\infty}^{+\infty} (g(\alpha u + \alpha\beta v) - g(\alpha u))^2 \cdot e^{-(u^2+v^2/2)} dudv \tag{4.6}$$

To show the effects of this method, let us consider the sensitivity of a threshold function, i.e., $A = +\infty$, $B = 1$ and $C = -1$. Eq. (4.5) can be calculated by (Yeung and Sun, 2002):

$$S_R(y_j^l) = \sqrt{\frac{4}{\pi} arctg \sqrt{\frac{\sigma_{\Delta x}^2}{\sigma_x^2} + \frac{\sigma_{\Delta w}^2}{\sigma_w^2} + \frac{\sigma_{\Delta x}^2}{\sigma_x^2} \cdot \frac{\sigma_{\Delta w}^2}{\sigma_w^2}}} \tag{4.7}$$

Comparing Eq. (4.5) with Eq. (4.7), we can see that, when $\frac{\sigma_{\Delta x}^2}{\sigma_x^2}$ and $\frac{\sigma_{\Delta w}^2}{\sigma_w^2}$ are very small, they are nearly the same. But when they are large, the results are different. For the threshold activation function, the output error variance cannot exceed 2. The result derived from parameterized antisymmetric function satisfies this, but Piché's fails. So the parameterized antisymmetric squashing function is more reasonable, and it is valid when the perturbation is either small or large.

4.3 Summary

For each antisymmetric squashing activation function, there exist three factors A, B and C, so that $g_{A,B,C}(x)$ can approximate in the sense of the L^2 norm. So we can use a vector (A, B, C) to represent an antisymmetric squashing activation function. That is to say that we can treat the antisymmetric squashing activation function as a parameter involved in the sensitivity of the MLP.

Chapter 5
Localized Generalization Error Model

The generalization error bounds found by current error models using the number of effective parameters of a classifier and the number of training samples are usually very loose. These bounds are intended for the entire input space. However, support vector machines, radial basis function neural networks and multilayer Perceptron neural networks are local learning machines for solving problems and thus treat unseen samples near the training samples as more important. In this chapter, we describe a localized generalization error model which bounds the generalization error from above within a neighborhood of the training samples using a stochastic sensitivity measure (Yeung et al., 2007 and Ng et al., 2007). This model is then used to develop an architecture selection technique for a classifier with maximal coverage of unseen samples by specifying a generalization error threshold.

5.1 Introduction

For a pattern classification problem, one builds a classifier f_θ to approximate or mimic the unknown input-output mapping function F, where θ is the set of parameters selected from a domain Λ. For the model presented here, the mean-square-error (MSE) is used to measure the difference between f_θ and F. The MSE is widely applied to training real-value output classifiers like neural networks, which classify a given sample by thresholding the real-value classifier output. The behavior of a classifier trained by minimizing an MSE, compared to the one trained by minimizing a classification error (0-1 loss function), is different. When two classifiers both yield the same, but very small percentage of training classification error, the one which yields a larger training MSE would produce indecisive outputs that deviate more from the target outputs. So a small change in the inputs may change the classification results (Anthony and Bartlett, 1999). This is not desirable and it indicates that the training classification error is not a good benchmark for the generalization capability of a classifier. Therefore, selecting a classifier using the training classification error or its bound may not be appropriate.

The classification error for the entire input space is defined as

$$R_{true} = \int_T \left(f_\theta\left(\mathbf{x}\right) - F\left(\mathbf{x}\right)\right)^2 p\left(\mathbf{x}\right) d\mathbf{x} \tag{5.1}$$

D.S. Yeung et al., *Sensitivity Analysis for Neural Networks*, Natural Computing Series, DOI 10.1007/978-3-642-02532-7_5, © Springer-Verlag Berlin Heidelberg 2010

where $\mathbf{x}$ denotes the input vector of a sample in the entire input space T, and $p(\mathbf{x})$ denotes the true unknown probability density function of the input $\mathbf{x}$.

Given a training dataset D containing N training input-output pairs $D = \{(\mathbf{x}_b, F(\mathbf{x}_b))\}_{b=1}^{N}$, a classifier f_θ could be constructed by minimizing the empirical risk (R_{emp}) over D, where

$$R_{emp} = \frac{1}{N}\sum_{b=1}^{N}\left(f_\theta(\mathbf{x}_b) - F(\mathbf{x}_b)\right)^2 \tag{5.2}$$

The ultimate goal of solving a pattern classification problem is to find the f_θ that is able to correctly classify future unseen samples (Haykin, 1999; Vapnik, 1998). The generalization error (R_{gen}) is defined as:

$$R_{gen} = \int_{T\backslash D}\left(f_\theta(\mathbf{x}) - F(\mathbf{x})\right)^2 p(\mathbf{x})d\mathbf{x} \tag{5.3}$$

Since both target outputs and distributions of the unseen samples are unknown, it is impossible to compute the R_{gen} directly. There are two major approaches to estimate the R_{gen}, namely, an analytical model and cross-validation (CV).

In general, analytical models cannot distinguish trained classifiers having the same number of effective parameters, but with different values of parameters. Thus they yield loose error bounds. The Akaike information criterion (AIC) (Akaike, 1974) only makes use of the number of effective parameters and the number of training samples. The Network Information Criterion (NIC) (Park et al., 2004) is an extension of the AIC for application in regularized neural networks. It defines classifier complexity by the trace of the second-order local derivatives of the network outputs with respect to connection weights. Unfortunately, current analytical models are only good for linear classifiers due to the singularity problem in their models. A major problem with these models is the difficulty of estimating the number of effective parameters of the classifier. This could be solved by using the Vapnik-Chervonenkis (VC) dimensions (Vapnik, 1998). However, only loose bounds of VC dimensions could be found for nonlinear classifiers, thus severely limiting the applicability of analytical models to nonlinear classifiers, except for the support vector machine (SVM).

Although CV uses true target outputs for unseen samples, it is time consuming for large datasets and, for k-fold CV and L choices of classifier parameters, kL classifiers must be trained. CV methods estimate the expected generalization error instead of its bound. Thus they cannot guarantee that the classifier finally constructed will have good generalization capability.

Many classifiers, e.g., SVM, multilayer Perceptrons (MLPNN) and radial basis function neural networks (RBFNNs) are local learning machines. An RBFNN, by its nature, learns classification locally and every hidden neuron captures local information of a particular region in the input space defined by the center and width of its Gaussian activation function (Moody and Darken, 1989). A training sample located far away from a hidden neuron's center does not affect the learning of this hidden neuron. An MLPNN learns decision boundaries for the classification problem in input space using the location of the training samples. However, as pointed out by Chakraborty and Pal (2001), the MLPNN output responses to unseen samples far away from the training samples are likely to be unreliable. So they proposed a learning algorithm which deactivates any MLPNN response to unseen samples much different from the training samples. This observation is further supported by the fact that, in many interesting industrial applications, such as aircraft detection in SAR images, character and fingerprint recognition (Jain and Venuri, 1999), and so on, the most significant unseen samples are expected to be similar to the training samples.

On the other hand, an RBFNN is one of the most widely applied neural networks for pattern classification, with its performance primarily determined by its architecture selection. Haykin (1999) summarizes several training algorithms for an RBFNN. For instance, a two-stage learning algorithm may be a quick way to train an RBFNN. The first, unsupervised stage is to select the center positions and widths for the RBF using self-organizing map or k-means clustering. The second, supervised stage computes the connection weights using the least mean square method or pseudoinverse technique. Some have proposed that all training samples be selected as centers (Haykin, 1999). Mao (2002) based the selection of centers on the separability of the datasets. The experimental results of Mao (2002) indicate that the choice of the number of hidden neurons indeed affects the generalization capability of the RBFNNs and that an increase in the number of hidden neurons does not necessarily lead to a decrease in testing error. The selection of the number of hidden neurons affects the selection of an RBFNN architecture and ad hoc choices or sequential search are frequently used.

In this chapter we describe a localized generalization error model R_{SM} using the stochastic sensitivity measure (ST-SM), which bounds the generalization error from above for unseen samples within a predefined neighborhood of the training samples (Yeung et al., 2007 and Ng et al., 2007). In addition, an architecture selection method based on the R_{SM} is proposed to find the maximal coverage classifier with its R_{SM} bounded by a preselected threshold. An RBFNN will be used to demonstrate the use of the R_{SM} and the architecture selection method. We introduce the localized generalization error model and its corresponding architecture selection method in Sects. 5.2 and 5.3 respectively.

5.2 The Localized Generalization Error Model

Two major concepts of the R_{SM}, the Q-neighborhood and the stochastic sensitivity measure, are introduced in Sects. 5.2.1 and 5.2.3 respectively. The derivation of the localized generalization error model is given in Sect. 5.2.2 and its characteristics are discussed in Sect. 5.2.4. Section 5.2.5 discusses the method to compare two classifiers using the localized generalization error model.

5.2.1 The Q-Neighborhood and Q-Union

For every sample $\mathbf{x}_b \in D$, one finds a set of samples $\mathbf{x}$ which fulfills $0 < |\Delta x_i| < Q$ $\forall i = 1 \cdots n$, where n denotes the number of input features, $\Delta \mathbf{x} = (\Delta x_1 \cdots \Delta x_n)' = \mathbf{x} - \mathbf{x}_b$ and Q is a given real number. In a pattern classification problem, one usually does not have any knowledge about the distribution of the true input space. Therefore, without any prior knowledge, every unseen sample has the same chance to appear. So, $\Delta \mathbf{x}$ may be considered as input perturbations that are random variables having zero-mean uniform distributions.

$$T_Q(\mathbf{x}_b) = \{\mathbf{x} \mid \mathbf{x} = \mathbf{x}_b + \Delta \mathbf{x}; |\Delta x_i| \le Q \quad \forall i = 1, \cdots, n\} \tag{5.4}$$

Then $T_Q(\mathbf{x}_b)$ defines a Q-neighborhood of the training sample $\mathbf{x}_b$ Let T_Q be the union of all $T_Q(\mathbf{x}_b)$ and call it the Q-Union. All samples in $T_Q(\mathbf{x}_b)$, except $\mathbf{x}_b$, are considered as unseen samples (i.e., $T_Q(\mathbf{x}_b)$ contains no training point other than $\mathbf{x}_b$). For $0 \le Q_1 \le \cdots \le Q_k \le \infty$, the following relationship holds:

$$D \subseteq T_{Q_1} \subseteq \cdots \subseteq T_{Q_k} \subseteq T \tag{5.5}$$

One should note that the shape of the Q-neighborhood is chosen to be a hypercube for ease of computation, but it could also be a hypersphere or other shape. Moreover, in the localized generalization error model, the unseen samples could be selected from a distribution other than a uniform one. Only the derivation of the ST-SM needs to be modified and the rest of the model will remain the same.

5.2.2 The Localized Generalization Error Bound

Instead of finding a bound for the generalization error for unseen samples in the entire input space T (R_{true}), we find a bound on R_{SM}, which is the error for unseen samples within T_Q only, i.e., the shaded area in Fig. 5.1. We ignore the generalization error for unseen samples that are located far away from training samples (R_{res} in Eq. (5.6)). Note that R_{res} decreases when Q increases.

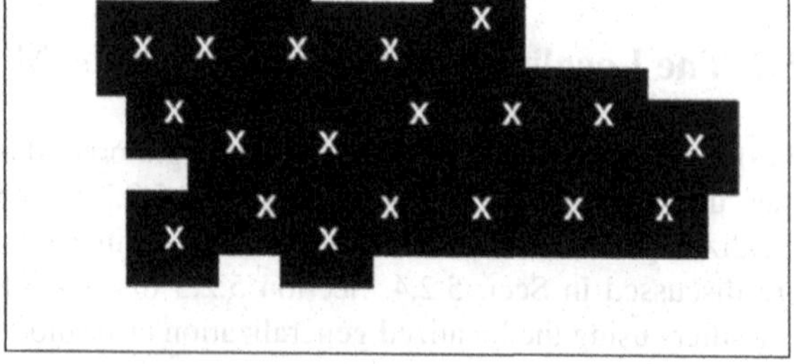

Fig. 5.1 An Illustration of Q-Union (T_Q) with 20 training samples. The Xs are training samples and any point in the shaded area is an unseen sample

$$R_{SM}(Q) = R_{true} - R_{res}(Q) = \int_{T_Q} (f_\theta(\mathbf{x}) - F(\mathbf{x}))^2 p(\mathbf{x})\, d\mathbf{x} \tag{5.6}$$

Let $\Delta y = f_\theta(\mathbf{x}) - f_\theta(\mathbf{x}_b)$, $err_\theta(\mathbf{x}_b) = f_\theta(\mathbf{x}_b) - F(\mathbf{x}_b)$; then $R_{emp} = (1/N) \sum_{b=1}^{N} (err_\theta(\mathbf{x}_b))^2$, $E_{T_Q}\left((\Delta y)^2\right) = \frac{1}{N} \sum_{b=1}^{N} \int_{T_Q(\mathbf{x}_b)} (\Delta y)^2 \frac{1}{(2Q)^n} d\mathbf{x}$, $\varepsilon = B\sqrt{\ln \eta / (-2N)}$; and let A, B and N be the difference between the maximum and minimum values of the target outputs, the maximum possible value of the MSE and the number of training samples, respectively. In this work, we assume that the range of the desired output, i.e., the range of F, is either known or assigned to a preselected value. Moreover, B is computable because the range of the network outputs will be known after the classifier is trained. In general, one expects that the error of unseen samples will be larger than the training error, so we assume that the average of errors of unseen samples in $T_Q(\mathbf{x}_b)$ will be larger than the training error of $\mathbf{x}_b$. By the Hoeffding inequality (Hoeffding, 1963), the average of the square errors of samples with the same population mean converges to the true mean with the following rate of convergence. With a probability of$(1 - \eta)$, we have (Yeung et al., 2007 and Ng et al., 2007):

$$R_{SM}(Q) \le \frac{1}{N} \sum_{b=1}^{N} \int_{T_Q(\mathbf{x}_b)} (f_\theta(\mathbf{x}) - F(\mathbf{x}))^2 \frac{1}{(2Q)^n} d\mathbf{x} + \varepsilon$$

$$= \frac{1}{N} \sum_{b=1}^{N} \int_{T_Q(\mathbf{x}_b)} (f_\theta(\mathbf{x}) - f_\theta(\mathbf{x}_b) + f_\theta(\mathbf{x}_b) - F(\mathbf{x}_b) + F(\mathbf{x}_b) - F(\mathbf{x}))^2 \frac{1}{(2Q)^n} d\mathbf{x} + \varepsilon$$

$$\le \frac{1}{N} \sum_{b=1}^{N} \int_{T_Q(\mathbf{x}_b)} \left((\Delta y)^2\right) \frac{1}{(2Q)^n} d\mathbf{x}$$

$$+ \frac{1}{N} \sum_{b=1}^{N} \int_{T_Q(\mathbf{x}_b)} \left((err_\theta(\mathbf{x}_b))^2\right) \frac{1}{(2Q)^n} d\mathbf{x} + \frac{1}{N} \sum_{b=1}^{N} \int_{T_Q(\mathbf{x}_b)} \left((F(\mathbf{x}_b) - F(\mathbf{x}))^2\right) \frac{1}{(2Q)^n} d\mathbf{x} + \varepsilon$$

$$+ 2\sqrt{\left(\frac{1}{N} \sum_{b=1}^{N} \int_{T_Q(\mathbf{x}_b)} \left((\Delta y)^2\right) \frac{1}{(2Q)^n} d\mathbf{x}\right) \left(\frac{1}{N} \sum_{b=1}^{N} \int_{T_Q(\mathbf{x}_b)} \left((err_\theta(\mathbf{x}_b))^2\right) \frac{1}{(2Q)^n} d\mathbf{x}\right)}$$

$$+ 2\sqrt{\left(\frac{1}{N} \sum_{b=1}^{N} \int_{T_Q(\mathbf{x}_b)} \left((err_\theta(\mathbf{x}_b))^2\right) \frac{1}{(2Q)^n} d\mathbf{x}\right) \left(\frac{1}{N} \sum_{b=1}^{N} \int_{T_Q(\mathbf{x}_b)} \left((F(\mathbf{x}_b) - F(\mathbf{x}))^2\right) \frac{1}{(2Q)^n} d\mathbf{x}\right)}$$

$$+2\sqrt{\left(\frac{1}{N}\sum_{b=1}^{N}\int_{T_Q(\mathbf{x}_b)}\left((\Delta y)^2\right)\frac{1}{(2Q)^n}d\mathbf{x}\right)\left(\frac{1}{N}\sum_{b=1}^{N}\int_{T_Q(\mathbf{x}_b)}\left((F(\mathbf{x}_b)-F(\mathbf{x}))^2\right)\frac{1}{(2Q)^n}d\mathbf{x}\right)}$$

$$\leq\left(\sqrt{R_{emp}}+\sqrt{E_{T_Q}\left((\Delta y)^2\right)}+A\right)^2+\varepsilon$$

$$=R^*_{SM}(Q) \tag{5.7}$$

Both A and ε are constants for a given training dataset when an upper bound of the classifier output values is preselected. The R^*_{SM} is an upper bound for the MSE of the trained classifier for unseen samples within the Q-union. This bound is better than those regression error bounds (based on AIC and VC-dimension) that are defined by using only the number of effective parameters and training samples, while ignoring statistical characteristics of a training dataset such as mean and variance. Moreover, those error bounds usually grow quickly with the increase of the number of effective parameters, e.g., number of hidden neurons in an RBFNN and VC-Dimension, while the R^*_{SM} grows much slower. The term $E_{T_Q}\left((\Delta y)^2\right)$ will be discussed in Sect. 5.2.3. Further discussion on the characteristics of the R^*_{SM} will be given in Sect. 5.2.4.

5.2.3 Stochastic Sensitivity Measure for RBFNN

The output perturbation (Δy) measures the network output difference between the training sample $(\mathbf{x}_b \in D)$ and the unseen sample in its Q-neighborhood $((\mathbf{x}_b + \Delta\mathbf{x}) \in T_Q(\mathbf{x}_b))$. Thus, the ST-SM measures the expectation of the squares of network output perturbations (Δy) between training samples in D and unseen samples in T_Q.

The Sensitivity Measure (SM) of a neural network (Ng and Yeung, 2002; Ng et al., 2007) gives quantified data on the change of network outputs with respect to change of network inputs. Intuitively, it measures how sensitive the network output is to the input change. Ng et al. (2007) allowed every input or weight to have its own mean and variance, and the input and weight perturbations are allowed to be arbitrary. Hence the perturbed samples ($\mathbf{x}$ in Sect. 5.2.1) can be considered as unseen samples around the training samples $(\mathbf{x}_b)$. Ng and Yeung (2002) developed an analytical formula of the ST-SM for a Gaussian activation function RBFNN that is independent of the number of training samples. We assume the inputs are independent and not identically distributed and weight perturbations are not considered in this chapter. So, every input feature has its own expectation μ_{x_i} and variance $\sigma^2_{x_i}$. The input perturbation of the ith input feature is a random variable having a uniform distribution with mean 0 and variance $\sigma^2_{\Delta x_i}$. The centers and widths of the hidden neurons are constant and the connection weights are fixed beforehand. An RBFNN could be described as

$$f_\theta(\mathbf{x}) = \sum_{j=1}^{M} w_j \exp\left(\frac{\left\|\mathbf{x}-\mathbf{u}_j\right\|^2}{-2v_j^2}\right) = \sum_{j=1}^{M} w_j \phi_j(\mathbf{x}) \tag{5.8}$$

where M, $\mathbf{u}$ and v_j denote the number of hidden neurons, the center and width of the jth RBFNN hidden neuron respectively, and w_j denotes the connection weight between the jth hidden neuron and its corresponding output neuron. Let $\varphi_j = (w_j)^2 \exp\left(\left(Var\left(\|\mathbf{x}-\mathbf{u}_j\|^2\right)/2v_j^4\right) - \left(E\left(\|\mathbf{x}-\mathbf{u}_j\|^2\right)/v_j^2\right)\right)$, $E\left(\|\mathbf{x}-\mathbf{u}_j\|^2\right) = \sum_{i=1}^{n}\left(\sigma_{x_i}^2 + \left(\mu_{x_i} - u_{ji}\right)^2\right)$, $\upsilon_j = \varphi_j\left(\sum_{i=1}^{n}\left(\sigma_{x_i}^2 + \left(\mu_{x_i} - u_{ji}\right)^2\right)/v_j^4\right)$, $\zeta_j = \varphi_j/v_j^4$,
$Var\left(\|\mathbf{x}-\mathbf{u}_j\|^2\right) = \sum_{i=1}^{n}\left(\begin{array}{l} E_D\left[\left(x_i-\mu_{x_i}\right)^4\right] - \left(\sigma_{x_i}^2\right)^2 + 4\sigma_{x_i}^2\left(\mu_{x_i}-u_{ji}\right)^2 \\ +4E_D\left[\left(x_i-\mu_{x_i}\right)^3\right]\left(\mu_{x_i}-u_{ji}\right)\end{array}\right)$,
$p(\Delta\mathbf{x})$ denotes the probability density function of the input perturbations and $p(\Delta\mathbf{x}) = 1/(2Q)^n$, n is the number of input features and u_{ji} denotes the ith input feature of the jth center of the hidden RBF neuron ($\mathbf{u}_j = \left(u_{j1},\cdots,u_{jn}\right)'$). For uniformly distributed input perturbations, we have $\sigma_{\Delta x_i}^2 = (2Q)^{2/12} = Q^{2/3}$. Theoretically, we do not restrict the distribution of the input perturbations as long as the variance of the input perturbation ($\sigma_{\Delta x_i}^2$) is finite. However, uniform distribution is assumed here because without any prior knowledge on the distribution of unseen samples around the training samples, we assume that all of them have an equal chance of occurrence.

By the law of large numbers, when the number of input features is not too small, $\phi_j(\mathbf{x})$ would have a lognormal distribution. So, the RBFNN ST-SM is given by:

$$E_{T_Q}\left((\Delta y)^2\right) = \frac{1}{N}\sum_{b=1}^{N}\int_{T_Q(\mathbf{x}_b)} \left(f_\theta(\mathbf{x}_b+\Delta\mathbf{x}) - f_\theta(\mathbf{x}_b)\right)^2 p(\Delta\mathbf{x})\, d\Delta\mathbf{x}$$

$$= \sum_{j=1}^{M}\left[\varphi_j\left(\begin{array}{l} \exp\left(\left(4\sum_{i=1}^{n}\sigma_{\Delta x_i}^2\left(\sigma_{x_i}^2+\left(\mu_{x_i}-u_{ji}\right)^2+0.2\sigma_{\Delta x_i}^2\right)\right)/\left(2v_j^4\right) - \left(2\sum_{i=1}^{n}\sigma_{\Delta x_i}^2\right)/\left(2v_j^2\right)\right) - \\ 2\exp\left(\left(\sum_{i=1}^{n}\sigma_{\Delta x_i}^2\left(\sigma_{x_i}^2+\left(\mu_{x_i}-u_{ji}\right)^2+0.2\sigma_{\Delta x_i}^2\right)\right)/2v_j^4 - \left(\sum_{i=1}^{n}\sigma_{\Delta x_i}^2\right)/2v_j^2\right) + 1 \end{array}\right)\right]$$

$$\approx \sum_{j=1}^{M}\varphi_j\left(\left(\sum_{i=1}^{n}\sigma_{\Delta x_i}^2\left(\sigma_{x_i}^2+\left(\mu_{x_i}-u_{ji}\right)^2+0.2\sigma_{\Delta x_i}^2\right)\right)/v_j^4\right)$$

$$= \frac{1}{3}Q^2\sum_{j=1}^{M}\upsilon_j + \frac{0.2}{9}Q^4 n\sum_{j=1}^{M}\zeta_j \tag{5.9}$$

5.2.4 Characteristics of the Error Bound

From Eq. (5.7), one may notice that the R^*_{SM} consists of three major components: training error (R_{emp}), ST-SM ($E_{T_Q}\left((\Delta y)^2\right)$) and the constants. The constants A and ε are preselected when the confidence of the bound $(1-\eta)$ and the training dataset is fixed. Moreover, the constant B in ε could be preselected when the classifier type is selected by fixing the maximum classifier output bound. ε is generally very small for large N. So, they will not affect the result of comparisons of the generalization capability between classifiers. In contrast, if the classifier could not generalize the training samples, one may not expect the classifier to have good generalization capability for future unseen samples. Thus the training error is one of the key components of the R^*_{SM}. Furthermore, the ST-SM term measures the output fluctuations of the classifier. A classifier having high output fluctuations yields a high ST-SM because its output varies dramatically when the input value changes. So, due to the classifier Bias/Variance Dilemma, a classifier yielding a good generalization capability should minimize both terms or achieve a good balance between the two (Geman and Bienenstock, 1992).

An interesting question is, can R^*_{SM} be an effective mechanism for studying a classifier's Bias/Variance Dilemma?

Limiting cases of $\mathbf{R^*_{SM}}$ (Q). Obviously, when $Q \to \infty$, $T_Q \to T$ and $Q \to 0$, $T_Q \to 0$. For$0 \leq Q_1 \leq \cdots \leq Q_k \leq (Q \to \infty)$, the relationship $D \subseteq T_{Q_1} \subseteq \cdots \subseteq T_{Q_k} \subseteq T$ holds. We further extend this relationship to:

$$\begin{aligned} &R_{emp}\left(\text{the limiting case of } R^*_{SM} \text{ with } Q \to 0\right) \leq R^*_{SM}(Q_1) \\ &\leq \cdots \leq R^*_{SM}(Q_k) \leq R_{true}\left(R^*_{SM} \; with \; Q \to \infty\right) \end{aligned} \tag{5.10}$$

This relationship shows that the limiting case of $R^*_{SM}(Q)$ with $Q \to \infty$ bounds from above the R_{true}. For $Q \to \infty$,

$$\begin{aligned} R_{true} &= \int_T (f_\theta(\mathbf{x}) - F(\mathbf{x}))^2 p(\mathbf{x})\, d\mathbf{x} \\ &= \int_{T_Q} (f_\theta(\mathbf{x}) - F(\mathbf{x}))^2 p(\mathbf{x})\, d\mathbf{x} \\ &= R_{SM}(Q) \leq R^*_{SM}(Q) \end{aligned} \tag{5.11}$$

Moreover, for the limiting case of $Q \to 0$, $R^*_{SM}(Q)$ bounds R_{emp} from above. When $Q \to 0$, $E_{T_Q}\left((\Delta y)^2\right)$ vanishes, and we have

$$R_{emp} \leq \left(\sqrt{R_{emp}} + A\right)^2 + \varepsilon \leq R^*_{SM}(Q) \tag{5.12}$$

R^*_{SM} for other classifiers. The R^*_{SM} as well as the R_{SM}, could be defined for any classifier trained with MSE. Examples include feedforward neural networks like multilayer Perceptron neural networks, support vector machines and recurrent neural networks such as Hopfield networks. The R^*_{SM} for other types of classifier could be defined by rederiving the ST-SM term for the particular type of classifier concerned.

Independence of training method. The R^*_{SM} is determined without regard of the training methods being used. Only the parameters of the finally trained classifier are used in the model. Hence the R^*_{SM} model could also be used to compare different training methods in terms of the generalization capability of the classifiers being built.

Time complexity. The ST-SM has a time complexity of O(Mn). The computational complexities of both the ST-SM and the R^*_{SM} are low and they do not depend on the number of training samples (N). However, similarly to all other architecture selection methods, R^*_{SM} requires a trained classifier and the time for architecture selections is dominated by the classifier training time. Therefore, the proposed architecture selection method may not have a large advantage in terms of speed in comparison to other architecture selection methods, except the two cross-validation-based methods.

Limitations of the localized generalization error model. The major limitation of the localized generalization error model is that it requires a function (i.e., f_θ) for the derivation of the ST-SM. Classifiers such as rule-based systems and decision trees may not be able to make use of this concept. It will be a challenging task to find ways to determine the R^*_{SM} for these classifiers.

Another limitation of the present localized generalization error model is due to the assumption that unseen samples are uniformly distributed. This assumption is reasonable when there is no a priori knowledge of the true distribution of the input space and hence every sample may have the same probability of occurrence. One would need to re-derive a new R^*_{SM} when a different distribution of the input space is assumed. We also remark that even if the distribution of the unseen samples is known, their true target outputs are still unknown and hence it will be difficult to judge how good the bound is. On the other hand, if both input and Q-Union distributions are the same, the localized generalization error model is expected to be a good estimate of the generalization error for the unseen samples, but this needs further investigation.

R^*_{SM} and regularization. From Eq. (5.9), one may notice that the connection weight magnitude (w_j^2 in the φ_j) is directly proportional to the ST-SM and thus the R^*_{SM}. This provides a theoretical justification that, by controlling the magnitude of the connection weights between hidden and output layers, one could reduce the generalization error which is essential to the regularization of neural network learning (Haykin, 1999).

Predicting unseen samples outside the Q-Union. In practice, some unseen samples may be located outside the Q-Union. This may be due to the fact that the Q value is too small and thus the Q-Union covers only very few unseen samples. However, expanding the Q value will lead to a larger R^*_{SM}, because more dissimilar unseen samples are included in the Q-Union, and a classifier with a very large $\mathrm{R}^*_{\mathrm{SM}}$ upper bound may not be meaningful. Furthermore, the $\mathrm{R}^*_{\mathrm{SM}}$ bounds from above the MSE of the unseen samples, and the MSE is an average of the errors. Thus, the $\mathrm{R}^*_{\mathrm{SM}}$ bound may still work well even if some of the unseen samples are located outside the Q-Union. When splitting the training and testing datasets, naturally some unseen testing samples may fall outside the Q-Union. However, our experiments show that the generalization capability of the RBFNNs selected by using the $\mathrm{R}^*_{\mathrm{SM}}$ is still the best in terms of testing accuracy when compared with other methods.

On the other hand, if there is a large portion of unseen samples located outside the Q-Union, i.e., dissimilar to the training samples, one may consider revising the training dataset to include more such samples and retrain the classifier. As mentioned before, classifiers may not be expected to classify unseen samples that are totally different from the training set.

5.2.5 *Comparing Two Classifiers Using the Error Bound*

One way to compare two classifiers is to fix the R^*_{SM} (Q) value and compare the difference in their Q values. The other way is to fix the Q value and compare the R^*_{SM} (Q) of the two classifiers.

Assume there are two classifiers, f_1 and f_2. There exists a Q_1 for f_1 yielding R^*_{SM} $(Q_1) = a$ and a Q_2 for f_2 yielding R^*_{SM} $(Q_2) = a$. If $Q_1 < Q_2$, then f_2 has a better generalization capability because f_2 covers more unseen samples, but still has the same generalization error upper bound. In other words, the architecture selection could be done by searching the functional space $\theta \in \Lambda$ and seeking the f_θ with the largest Q producing R^*_{SM} $(Q) = a$. This is an important property of R^*_{SM} (Q) and we will make use of it to find the optimal classifier with the largest coverage in the next section.

On the other hand, one could compare the two classifiers, f_1 and f_2, based on their R^*_{SM} (Q) computed using the same Q value. The classifier with the lower R^*_{SM} (Q) is expected to have a better generalization capability. All other comparison methods in which neither Q nor R^*_{SM} (Q) is fixed may not be meaningful.

5.3 Architecture Selection Using the Error Bound

The selection of the number of hidden neurons in the RBFNN is usually done by Sequential Learning (Huang et al., 2005) or by ad hoc choice. The Sequential Learning technique only makes use of the training error to determine the number of hidden neurons, without any reference to the generalization capability. Moreover, Huang et al. (2005) and Liang et al. (2006) assume the classifier does not have prior knowledge about the number of training samples while Kaminski and Strumillo (1997) and Gomm and Yu (2000) assume that it does. For ease of comparison with other architecture selection methods, we assume that the number of training samples in our experiments is known to the classifier. In this section, we describe a new technique based on R^*_{SM} to find the optimal number of hidden neurons that makes use of the generalization capability of the RBFNN.

For any given threshold a on the generalization error bound (R^*_{SM}), the localized generalization error model allows us to find the best classifier by maximizing Q, assuming that the MSE of all samples within the Q-Union is smaller than a. One can formulate the architecture selection problem as a Maximal Coverage Classification problem with Selected Generalization error bound (MC^{2}SG), i.e.,

$$\max_{\theta \in \Lambda} \quad Q \quad \text{subject to} \quad R^*_{SM}(Q) \leq a \tag{5.13}$$

In the RBFNN training algorithm presented in Sect. 5.3.2, once the number of hidden neurons is fixed, the center positions and widths could be estimated by any automatic clustering algorithm such as k-means clustering, a self-organizing map or hierarchical clustering. So we only need to concentrate on the problem of determining the number of hidden neurons. This means that $\theta = M$ and $\Lambda = \{1, 2 \ldots \mathrm{N}\}$ because it is not reasonable to have the number of hidden neurons larger than the number of training samples.

Eq. (5.13) represents a two-dimensional optimization problem. The first dimension is the number of hidden neurons (θ) and the second dimension is the Q for a fixed θ. For every fixed θ and Q, we can determine an $R^*_{SM}(Q)$. The two parameters, θ and Q, are independent. Furthermore, by substituting Eq. (5.9) into Eq. (5.7), with probability $(1-\eta)$ we have the R^*_{SM} for an RBFNN as follows,

$$R^*_{SM}(Q) \approx \left(\sqrt{\frac{1}{3}Q^2 \sum_{j=1}^{M} \upsilon_j + \frac{0.2}{9} Q^4 n \sum_{j=1}^{M} \zeta_j} + \sqrt{R_{emp}} + A \right)^2 + \varepsilon \tag{5.14}$$

Let $R^*_{SM}(Q) = a$. For every θ, let the Q that satisfies $R^*_{SM}(Q) = a$ be Q^*, where a is a constant real number. $R^*_{SM}(Q) = a$ exists because the second-order derivative of Eq. (5.14) is positive. We could solve Eq. (5.14) as follows,

$$Q^4 \frac{0.2}{3} n \sum_{j=1}^{M} \zeta_j + Q^2 \sum_{j=1}^{M} \upsilon_j - 3\left(\sqrt{a-\varepsilon} - \sqrt{R_{emp}} - A\right)^2 = 0 \tag{5.15}$$

Equation (5.15) could be solved by the quadratic equation and two solutions will be found for Q^2. For $a \geq \varepsilon$ (ε is usually a very small constant when the number of samples is large), there will be one positive and one negative real solution for Q^2 because in Eq. (5.15), the coefficients for the terms Q^4 and Q^2 are positive, but the constant term is negative. Also, there will be two real and two imaginary solutions for Q. Let Q^* be the only positive real solution among the four. Note that Q is defined to be the width of the Q-neighborhood and as such it must be a nonnegative real number.

$$h\left(M,Q^*\right) = \begin{cases} 0 & R_{emp} \geq a \\ Q^* & else \end{cases} \tag{5.16}$$

So, for RBFNN architecture selection, Eq. (5.13) is equivalent to:

$$\max_{M \in \Lambda} \quad h\left(M,Q^*\right) \tag{5.17}$$

5.3.1 *Parameters for MC²SG*

In Eq. (5.7), the differences between the maximum and minimum values of target outputs (A) and the number of training samples (N) are fixed for a given training dataset, and the maximum possible value of the MSE and the confidence level of the R^*_{SM} bound, namely, B and η, could also be selected before any classifier training.

In a K-class classification problem, one may select $F(\mathbf{x}) \in \{(k_1,k_2 \cdots k_K)\}$ where $k_i \in \{0,1\}$ and $\sum_{i=1}^{K} k_i = 1$. $k_i = 1$ if the sample belongs to the ith class, and one minimizes the MSE of all the RBFNN outputs simultaneously. All the $F(\mathbf{x}), f_\theta(\mathbf{x})$ and R^*_{SM} are vectors and thus the sum of R^*_{SM} values of all the K RBFNN outputs are minimized in the MC^{2}SG. One may notice that the minimization of the sum of the R^*_{SM} values of all the outputs is equivalent to the minimization of the average of them. However, the average of the R^*_{SM} values may provide a better interpretation and its range is not affected by the value K.

The determination of the constant a is made according to the classifier's output schemes for classification. For instance, if class outputs are different by 1, then a may be selected as 0.25 because a sample is misclassified if the square of its deviation from the target output is larger than 0.25. From Eq. (5.16), one may notice that the smaller the values of a is, the larger the number of hidden neurons selected by the MC^{2}SG. This is because a larger number of hidden neurons is needed to reduce the training error. On the other hand, if the a value is selected to be larger than 0.25, the effect on the architecture selection will be insignificant. Experimental results show that the RBFNNs yielding training error larger than 0.25 will not yield a good generalization capability, i.e., will yield poor testing accuracies (Yeung et al., 2007).

5.3.2 *RBFNN Architecture Selection Algorithm for MC²SG*

The solution of Eq. (5.17) is realized using the following selection algorithm. The MC^{2}SG is independent of any training algorithm of the RBFNN. Steps 2 and 3 represent unsupervised learning to find the centers and widths of the RBFs in the hidden neurons and Step 4 finds the least-square solution of the connection weights by making use of the linear relationship, shown in Eq. (5.8), between the hidden neuron outputs and RBFNN outputs. Other training algorithms could be adopted in Steps 2, 3 and 4 in the following Architecture Selection Algorithm.

The Architecture Selection Algorithm of the MC^{2}SG:

1. Start with $M = 1$ (M denotes the number of hidden neurons),
2. Execute k-means clustering algorithm to find the centers for the M hidden neurons,
3. For each of the M RBF hidden neurons, select its width value to be the distance between that center and its nearest hidden neuron,
4. Compute the connection weights using a pseudoinverse method,
5. Compute the Q-value for the current RBFNN using Eq. (5.16),
6. If the stopping criterion is not fulfilled, $M = M + 1$ and go to Step (2).

The stopping criterion could be selected as "M is equal to the number of training samples" and this will allow the MC^{2}SG to search for all possible hidden neurons. However, it is computationally prohibitive for large datasets and thus we will discuss a heuristic stopping criterion in Sect. 5.3.3. Moreover, a constructive approach is employed here because it is more efficient to start the search with one hidden neuron, and with each iteration add one hidden neuron.

5.3.3 *A Heuristic Method to Reduce the Computational Time for MC^{2}SG*

As in the other methods, $h(M, Q^*)$ is generally not differentiable with respect to M (not a smooth function). One must try out all possible M values in order to find the optimal solution. Our experimental results show that $h(M, Q^*)$ drops to 0 when the classifier becomes too complex, i.e., M is too large. Heuristically an early stop could be made to reduce the number of classifier training iterations when Q approaches 0. In our experiments, we stop the search when the Q values drop below a threshold. In fact, Q does not increase significantly after it drops below 10% of the maximum value of Q being found, and thus this is used as the threshold to speed up the MC^{2}SG.

5.4 Summary

In this chapter, we described a new generalization error model based on the localized generalization error. R^*_{SM} bounds the generalization error from above for unseen samples within the Q-neighborhoods of the training samples. Moreover, an architecture selection method, namely MC^{2}SG based on the R^*_{SM}, is proposed to find the RBFNN classifier that has the largest coverage of unseen samples, while it's R^*_{SM} is still less than a preselected threshold. The R^*_{SM} was shown to be a generalization of R_{true}.

This chapter has demonstrated the use of the MC^2SG to find the number of hidden neurons for an RBFNN, while the values of other parameters were found using existing methods. A possible extension of our result is to find the values of other RBFNN parameters, e.g., center positions, widths and connection weights, via an optimization of R_{SM}^* because these parameters are also coefficients of the R_{SM}^*. However, the trade-off between the optimality of the solution and time complexity will be an important consideration.

Chapter 6
Critical Vector Learning for RBF Networks

One of the most popular neural network models, the radial basis function (RBF) network attracts a lot of attention due to its improved approximation ability as well as the construction of its architecture. Bishop (1991) concluded that an RBF network can provide a fast, linear algorithm capable of representing complex nonlinear mappings. Park and Sandberg (1993) further showed that an RBF network can approximate any regular function. In a statistical sense, the approximation ability is a special case of statistical consistency. Hence, Xu et al. (1994) presented upper bounds for the convergence rates of the approximation error of RBF networks, and constructively proved the existence of a consistent point-wise estimator for RBF networks. Their results can be a guide to optimize the construction of an RBF network, which includes the determination of the total number of radial basis functions along with their centers and widths. This is an important problem to address because the performance and training of an RBF network depend very much on these parameters.

6.1 Related Work

There are three ways to construct an RBF network, namely, clustering, pruning and critical vector learning. Bishop (1991) and Xu (1998) followed the clustering method, in which the training examples are grouped and then each neuron is assigned to a cluster. The pruning method, such as Chen et al. (1991) and Mao (2002), creates a neuron for each training example and then prunes the hidden neurons by example selection. The critical vector learning method, outlined by Schölkopf et al. (1997) constructs an RBF with the critical vectors, rather than cluster centers.

Moody and Darken (1989) located an optimal set of centers using both the *k-means* clustering algorithm and learning vector quantization. The drawback of this method is that it considers only the distribution of the training inputs, yet the output values influence the positioning of the centers. Bishop (1991) introduced the Expectation-Maximization (EM) algorithm to optimize the cluster centers in two steps: obtaining of initial centers by clustering and optimization of the basis

D.S. Yeung et al., *Sensitivity Analysis for Neural Networks*, Natural Computing Series, DOI 10.1007/978-3-642-02532-7_6, functions by applying the EM algorithm. Such a treatment actually does not perform maximum likelihood learning but a suboptimal approximation. Xu (1998) extended the model for a mixture of experts to estimate basis functions, output neurons and the number of basis functions all together. The maximum likelihood learning and regularization mechanism can be further unified to his established Bayesian Ying Yang (BYY) learning framework (Xu, 2004a, 2004b, 2004c), in which any problem can be decomposed into Ying space or an invisible domain (e.g., the hidden neurons in RBFs), and Yang space or a visible domain (e.g., the training examples in RBFs), and the invisible/unknown parameters can be estimated through harmony learning between these two domains.

Chen et al. (1991) proposed orthogonal least square (OLS) learning to determine the optimal centers. The OLS combines the orthogonal transform with the forward regression procedure to select model terms from a large candidate term set. The advantage of employing an orthogonal transform is that the responses of the hidden layer neurons are decorrelated so that the contribution of individual candidate neurons to the approximation error reduction can be evaluated independently. However, the original OLS learning algorithm lacks generalization and global optimization abilities. Mao (2002) employed OLS to decouple the correlations among the responses of the hidden units so that the class separability provided by individual RBF neurons can be evaluated independently. This method can select a parsimonious network architecture as well as centers providing large class separation.

The common feature of all the above methods is that the radial basis function centers are a set of the optimal cluster centers of the training examples. Schölkopf et al. (1997) calculated *support vectors* using a support vector machine (SVM), and then used these support vectors as radial basis function centers. Their experimental results showed that the support-vector-based RBF outperforms conventional RBFs. Although the motivation of these researchers was to demonstrate the superior performance of a full support vector machine over either conventional or support-vector-based RBFs, their idea of *critical* vector learning is worth borrowing.

This chapter describes a novel approach to determining the centers of RBF networks based on sensitivity analysis.

6.2 Construction of RBF Networks with Sensitivity Analysis

An RBF classifier is a three-layer neural network model, in which an N-dimensional input vector $\mathbf{x}= (x_1\ x_2\ \ldots\ x_N)$ is broadcast to each of K neurons in the hidden layer. Each hidden neuron produces an activation function, typically a Gaussian kernel:

$$h_i = \exp\left(-\frac{\|\mathbf{x}-\mathbf{c}_i\|^2}{2\sigma_i^2}\right),\quad i = 1,2,\ldots,K \tag{6.1}$$

where $\mathbf{c}_i$ and σ_i^2 are the center and width of the Gaussian basis function of the i th hidden unit, respectively. The units in the output layer have interconnections with all the hidden units. The j th output neuron has the form:

$$f_j(\mathbf{x}) = \mathbf{w}_j\mathbf{h} = \sum_{i=1}^{K} w_{ij} \exp\left(-\frac{\|\mathbf{x}-\mathbf{c}_i\|^2}{2\sigma_i^2}\right) \tag{6.2}$$

where $\mathbf{h} = (h_1\ h_2 \ldots h_K)$ is the input vector from the hidden layer, and the w_{ij} is the interconnection weight between the j th output neuron and the i th hidden neuron.

In this section, the RBF classifier's sensitivity is defined as the mathematical expectation of the square of output deviations caused by the perturbation of RBF centers. An algorithm will be given that can be used to select critical vectors.

6.2.1 RBF Classifiers' Sensitivity to the Kernel Function Centers

We use symbols $\hat{\mathbf{c}}_i$ and $\hat{\sigma}_i$ to denote the values of center and width of the i th hidden neuron under a perturbation. Then the deviation resulting from this perturbation is:

$$\Delta y_j = \hat{\mathbf{w}}_j\hat{\mathbf{h}} - \mathbf{w}_j\mathbf{h} = \sum_{i=1}^{K} \hat{w}_{ij} \exp\left(-\frac{\left\|\mathbf{x}-\hat{\mathbf{c}}_i\right\|^2}{2\hat{\sigma}_i^2}\right) - \sum_{i=1}^{K} w_{ij} \exp\left(-\frac{\|\mathbf{x}-\mathbf{c}_i\|^2}{2\sigma_i^2}\right) \tag{6.3}$$

Here, $\hat{\mathbf{c}}_i = \mathbf{c}_i + \Delta\mathbf{c}_i$ are the centers deviated from the centers under the perturbations, and the interconnection weights under the perturbations are $\hat{\mathbf{w}}_j = \mathbf{w}_j + \Delta\mathbf{w}_j$, where $\mathbf{w}_\mathrm{j}$ can be calculated using a pseudo-matrix inversion, or data training. Although there are ways to specify RBF widths, such as the method of Xu et al. (2004), the most common method for selecting RBF widths is to make all of them equal to a constant value depending on the prior knowledge of the given application. With predefined RBF widths, we just focus on the perturbations on the centers and their interconnection weights in this chapter. The perturbation on the i th RBF center and the weights connected to the j th output, $\Delta\mathbf{c}_i$ and $\Delta\mathbf{w}_j$, can be generated following a Gaussian distribution with 0 means, variances $\sigma_{\mathbf{c}_i}$and $\sigma_{\mathbf{w}_j}$, respectively:

$$\begin{aligned} p(\Delta\mathbf{c}_i) &= \frac{1}{\left(\sqrt{2\pi\sigma_{\mathbf{c}_i}}\right)^N} \exp\left(-\frac{\Delta\mathbf{c}_i^T\Delta\mathbf{c}_i}{2\sigma_{\mathbf{c}_i}^2}\right) \\ p(\Delta\mathbf{w}_j) &= \frac{1}{\left(\sqrt{2\pi\sigma_{\mathbf{w}_j}}\right)^K} \exp\left(-\frac{\Delta\mathbf{w}_j^T\Delta\mathbf{w}_j}{2\sigma_{\mathbf{w}_j}^2}\right) \end{aligned} \tag{6.4}$$

where N is the dimension of the input $\mathbf{x}$, and K is the number of RBF centers.

The RBF centers will be selected recursively in the next subsection. To make the sensitivity analysis cater to the construction of RBF networks, a recursive definition of sensitivity is given below. At the K th time, suppose there are a number K–1 of RBF centers fixed already, and the newcomer $\mathbf{c}_i$ is observed. Hence, the j th output neuron's sensitivity to the current number K of RBF centers is defined as the mathematical expectation of $(\Delta y_j)^2$ (square of output deviations caused by the perturbations of RBF centers) with respect to all $\Delta \mathbf{c}_i$ and the training example set $D = \{\mathbf{x}_l\}_{l=1}^{L}$, which is expressed as

$$
\begin{aligned}
S_j^{(K)} &= E[(\Delta y_j)^2] = \\
&\int p(\Delta \mathbf{w}) p(\Delta \mathbf{c}) \sum_{\mathbf{x}_l \in D} \sum_{m,n=1}^{K} \hat{w}_{mj} \hat{w}_{nj} \exp\left(-\frac{\|\mathbf{x}_l - \mathbf{c}_m - \Delta \mathbf{c}_m\|^2}{2\sigma_m^2} - \frac{\|\mathbf{x}_l - \mathbf{c}_n - \Delta \mathbf{c}_n\|^2}{2\sigma_n^2}\right) d\Delta \mathbf{c} d\Delta \mathbf{w} \\
&-2 \int p(\Delta \mathbf{w}) p(\Delta \mathbf{c}) \sum_{\mathbf{x}_l \in D} \sum_{m,n=1}^{K} \hat{w}_{mj} w_{nj} \exp\left(-\frac{\|\mathbf{x}_l - \mathbf{c}_m - \Delta \mathbf{c}_m\|^2}{2\sigma_m^2} - \frac{\|\mathbf{x}_l - \mathbf{c}_n - \Delta \mathbf{c}_n\|^2}{2\sigma_n^2}\right) \\
&\qquad d\Delta \mathbf{c} d\Delta \mathbf{w} \\
&+ \int p(\Delta \mathbf{w}) p(\Delta \mathbf{c}) \sum_{\mathbf{x}_l \in D} \sum_{m,n=1}^{K} w_{mj} w_{nj} \exp\left(-\frac{\|\mathbf{x}_l - \mathbf{c}_m - \Delta \mathbf{c}_m\|^2}{2\sigma_m^2} - \frac{\|\mathbf{x}_l - \mathbf{c}_n - \Delta \mathbf{c}_n\|^2}{2\sigma_n^2}\right) d\Delta \mathbf{c} d\Delta \mathbf{w} \\
&= I_1 - 2I_2 + I_3
\end{aligned}
\tag{6.5}
$$

where I_1, I_2 and I_3 are figured out by integrating over $\Delta \mathbf{c}$ and $\Delta \mathbf{w}$, so $I_1=$

$$
\begin{aligned}
&\sum_{\mathbf{x}_l \in D} \sum_{m,n=1; m \neq n}^{K} w_{mj} w_{nj} \frac{\left(\sqrt{\sigma_m^2 \sigma_n^2}\right)^N}{\left(\sqrt{(\sigma_m^2 + \sigma_{\mathbf{c}m}^2)(\sigma_n^2 + \sigma_{\mathbf{c}n}^2)}\right)^N} \exp\left(-\frac{\|\mathbf{x}_l - \mathbf{c}_m\|^2}{2(\sigma_m^2 + \sigma_{\mathbf{c}m}^2)} - \frac{\|\mathbf{x}_l - \mathbf{c}_n\|^2}{2(\sigma_n^2 + \sigma_{\mathbf{c}n}^2)}\right), \\
&+ \sum_{\mathbf{x}_l \in D} \sum_{m=1}^{K} (w_{mj}^2 + \sigma_{\mathbf{w}_j}^2) \frac{\left(\sqrt{\sigma_m^2}\right)^N}{\left(\sqrt{(\sigma_m^2 + 2\sigma_{\mathbf{c}m}^2)}\right)^N} \exp\left(-\frac{\|\mathbf{x}_l - \mathbf{c}_m\|^2}{(\sigma_m^2 + 2\sigma_{\mathbf{c}m}^2)}\right)
\end{aligned}
$$

and similarly, we have

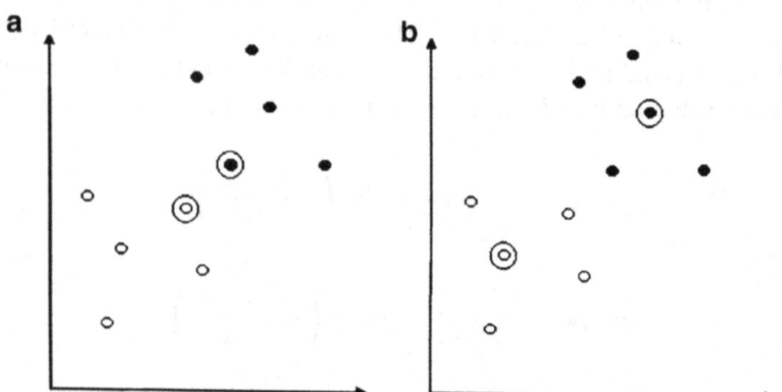

Fig. 6.1 Illustration of the difference between sensitivity-based and conventional RBF classifiers. The circles and balls represent data points of two classes respectively, and RBF centers are indicated by extra circles. (a) Sensitivity-based RBF, centers being the vectors sensitive to the classification. (b) Conventional RBF, centers being the cluster centroids

$$I_2 = \sum_{\mathbf{x}_l \in D} \sum_{m,n=1}^{K} w_{mj} w_{nj} \frac{\left(\sqrt{\sigma_m^2}\right)^N}{\left(\sqrt{(\sigma_m^2 + \sigma_{\mathbf{c}m}^2)}\right)^N} \exp\left(-\frac{\|\mathbf{x}_l - \mathbf{c}_m\|^2}{2(\sigma_m^2 + \sigma_{\mathbf{c}m}^2)} - \frac{\|\mathbf{x}_l - \mathbf{c}_n\|^2}{2\sigma_n^2}\right),$$

$$I_3 = \sum_{\mathbf{x}_l \in D} \sum_{m,n=1}^{K} w_{mj} w_{nj} \exp\left(-\frac{\|\mathbf{x}_l - \mathbf{c}_m\|^2}{2\sigma_m^2} - \frac{\|\mathbf{x}_l - \mathbf{c}_n\|^2}{2\sigma_n^2}\right).$$

The difference between the sensitivity-based and the conventional RBF networks can be illustrated as in Fig. 6.1.

6.2.2 Orthogonal Least Square Transform

The most critical vectors can be found by Eq. (6.5). However, the RBF centers cannot be determined by a sensitivity measure only, because some of the critical vectors may be correlated. The OLS (Chen et al., 1991) can alleviate this problem of redundant or correlated centers.

Let $\mathbf{Y} = (y_1, y_2 \ldots y_\mathrm{L})^\mathrm{T}$ be the output matrix corresponding to the number L of all training examples, y_i ($i = 1, 2, \ldots, L$) an M-dimensional vector denoting the number M of output units. We have

$$\mathbf{Y} = \mathbf{HW} = (\mathbf{QA})\,\mathbf{W} \tag{6.6}$$

where $\mathbf{Y}$, $\mathbf{H}$, $\mathbf{W}$ are $L \times M$, $L \times L$, and $L \times M$ matrices, respectively. The selection of RBF centers is equivalent to the selection of the most critical columns of H. The matrix H can be decomposed into QA, where Q is an $L \times L$ matrix with orthogonal columns $[\mathrm{q}_1, \mathrm{q}_2 \ldots \mathrm{q}_\mathrm{L}]$, and A is an $L \times L$ upper triangular matrix as follows:

$$\mathbf{H} = \begin{bmatrix} h_{11} & h_{12} & \cdots & h_{1L} \\ h_{12} & h_{12} & & \vdots \\ \vdots & & & \vdots \\ \vdots & & & \\ h_{1L} & \cdots & \cdots & h_{LL} \end{bmatrix}, \quad \mathbf{A} = \begin{bmatrix} 1 & a_{12} & \cdots & \cdots & a_{1L} \\ 0 & 1 & a_{23} & & \vdots \\ \vdots & 0 & \ddots & & \vdots \\ \vdots & & 0 & 1 & a_{(L-1)L} \\ 0 & \cdots & \cdots & 0 & 1 \end{bmatrix}.$$

Only one column of H is orthogonalized in each iteration. At the K th iteration, one column is made orthogonal to each of the K–1 previously orthogonalized columns. The computational procedure can be represented as follows (Chen et al. 1991):

$$\begin{cases} \mathbf{q}_1 = \mathbf{h}_1, \\ a_{ik} = \dfrac{\mathbf{q}_i^T \mathbf{h}_K}{\|\mathbf{q}_i\|}, \qquad 1 \le i < K \\ \mathbf{q}_K = \mathbf{h}_K - \sum\limits_{i=1}^{K-1} a_{iK} \mathbf{h}_i \end{cases} \tag{6.8}$$

then the RBF centers are selected by sorting these columns.

6.2.3 *Critical Vector Selection*

Let $\mathbf{S}^{(K)}(\mathbf{c}_i)$ denote the sensitivity of the previous $(K-1)$ RBF centers and a candidate RBF center c_i which corresponds to q_i at the K th time, where $1 \le i \le L$.

Substitute every interconnection weight in Eqs. (6.3) and (6.5):

$$w_{ij}^{(K)} = \sum_{l=1}^{L} a_{li} \cdot w_{ij}, \tag{6.9}$$

and calculate the sensitivity for all the possible K-center RBF networks. Let $\mathbf{Q}^{(K)}$ denote the orthogonal matrix at the K th time; then the columns in $\mathbf{Q}^{(K)}$ are sorted in the order:

$$\left\| \mathbf{S}^{(K)}(\mathbf{c}_1) \right\| \ge \left\| \mathbf{S}^{(K)}(\mathbf{c}_2) \right\| \ge \cdots \ge \left\| \mathbf{S}^{(K)}(\mathbf{c}_L) \right\| . \tag{6.10}$$

A formal statement of the algorithm for the selection of critical vectors is given as follows:

STEP 1. Initialization. Form the matrix H in Eq. (6.7) with the RBF function responses of all the training examples.

STEP 2. First critical vector neuron selection. Calculate the sensitivity of each column of H with Eq. (6.5). The column that provides the maximum sensitivity is selected as the first column of matrix $\mathbf{Q}^{(1)}$. Calculate the classification error $\mathbf{Err}^{(1)}$ with the selected RBF center. Let $K=2$.

STEP 3. Orthogonalization and critical vector selection. Orthogonalize all remaining columns of $\mathbf{H}$ with all the columns of $\mathbf{Q}^{(K-1)}$ using Eq. (6.8).

STEP 4. Each training example $\mathbf{c}_i$ is a candidate for the K th RBF center, which corresponds to the orthogonalized column $\mathbf{q}_i$, $(K \le i \le L)$. Calculate interconnection weights using pseudo-matrix inversion and compute the sensitivity of the previous $(K-1)$ RBF centers with each candidate center $\mathbf{S}^{(K)}(\mathbf{c}_i)$ with the weights updated by Eq. (6.9). Sort the columns in $\mathbf{Q}^{(K)}$ with Eq. (6.10), and the one yielding the maximum sensitivity is selected as the K th column of $\mathbf{Q}^{(K)}$. Calculate the classification error $\mathbf{Err}^{(K)}$ with the selected RBF centers.

STEP 5. If $(\mathbf{Err}^{(K)} - \mathbf{Err}^{(K-1)})$ is smaller than a predefined threshold, go to STEP 7.

STEP 6. K++, go to STEP 3.

STEP 7. End.

The critical vectors corresponding to the first K columns in $\mathbf{Q}^{(K)}$ will be selected as hidden neurons.

6.3 Summary

The conventional approach to constructing an RBF network is to search for the optimal cluster centers among the training examples. This chapter introduced a novel approach to RBF construction that uses critical vectors selected by sensitivity analysis. Sensitivity is defined as the expectation of the square of output deviations caused by the perturbation of RBF centers. In training, the orthogonal least square method incorporated with a sensitivity measure is employed to search for the optimal critical vectors. In classification, the selected critical vectors will take on the role of the RBF centers. The detailed experimental results can be seen in Shi, Yeung and Gao (2005), which shows that this sensitivity-based RBF classifier performs better than the conventional RBFs and C4.5. The sensitivity-based RBF can achieve the same level of accuracy as an SVM, but strikes a balance between critical vector learning and robustness.

Chapter 7
Sensitivity Analysis of Prior Knowledge[1]

The paradigm of Knowledge-Based Neurocomputing (Cloete et al., 2000b) addresses the encoding, extraction and refinement of symbolic knowledge in a neurocomputing paradigm. Prior symbolic knowledge derived outside of neural networks can be encoded in neural network form, and then further trained. Classification rules of various forms (Cloete, 1996 and 2000) are most often used as prior symbolic knowledge, and then transformed into a neural network that produces the same classification. In the transformation process, one would like to retain the flexibility in the network for further training, and encode the knowledge in such a way that it is not destroyed by further training, but can be revised. In addition, one would like the possibility to improve the classification by discovering new rules if needed.

These goals are conflicting, since the neural network can be encoded in such a way that it almost exactly represents the classification decisions, but then training with backpropagation is very difficult. This encoding provides an inductive bias, the choice of which has largely been an experimental procedure, except for the method proposed by Omlin et al. (2003) and Snyders et al. (2000).

This chapter uses sensitivity analysis methods (Engelbrecht et al., 1995; Zurada et al., 1994) to investigate the inductive bias of the encoding method for the prior knowledge. The chapter is organized as follows. The next section introduces the neural network encoding method, and in Sect. 7.2 the computation of the inductive bias is given. In Sect. 7.3 we derive the sensitivity measures, and in Sect. 7.4 we demonstrate the experimental results. The outcomes and conclusions are discussed in Sect. 7.5.

[1]© IEEE. Reprinted, with permission, from Cloete, I., Snyders, S., Yeung, D. S. and Wang, X. (2004). Sensitivity analysis of prior knowledge in knowledge-based neurocomputing. In *Proceedings of 2004 International Conference on Machine Learning and Cybernetics*, 7:4174–4181.

D.S. Yeung et al., *Sensitivity Analysis for Neural Networks*, Natural Computing Series,

DOI 10.1007/978-3-642-02532-7_7,

7.1 KBANNs

Several methods for encoding rules have been proposed in the literature (Cloete, 1996 and 2000). We use the method of Towell et al. (1994) to illustrate how Horn clauses can be encoded into feedforward networks. Some other methods only differ in the way neuron inputs are combined, e.g., Lacher et al. (1992). The construction of an initial network is based on the correspondence between entities of the knowledge base and neural networks, respectively. Supporting facts translate into input neurons, intermediate conclusions are modeled as hidden neurons, output neurons represent final conclusions; dependencies are expressed as weighted connections between neurons. See Fig. 7.1. The neuron outputs are computed by a sigmoidal function which takes a weighted sum of inputs as its argument.

Given a set of if-then rules (Fig. 7.1a), disjunctive rules are rewritten as follows: The consequent of each rule becomes the consequent of a single antecedent; it in turn becomes the consequent of the original rule (Fig. 7.1b). This rewriting step is necessary in order to prevent combinations of antecedents from activating a neuron when the corresponding conclusion cannot be drawn from such combinations. These rules are then mapped into a network topology as shown in Fig. 7.1c. A neuron is connected via weight H to a neuron in a higher level if that neuron corresponds to an antecedent of the corresponding conclusion. The weight of that connection is $+H$ if the antecedent is positive; otherwise, the weight is programmed to $-H$. For conjunctive rules, the neuron bias[2] of the corresponding consequent is set to $-(P-0.5)H$ where P is the number of positive antecedents; for disjunctive rules, the neuron bias is set to $-H/2$. This guarantees that neurons have a high output when all or any one of their antecedents have a high output for conjunctive and disjunctive rules, respectively. If the given initial domain theory is incomplete or incorrect, a network may be supplemented with additional neurons and weights which correspond to rules still to be learned from data (Fig. 7.1d).

The dynamics of a typical knowledge-based feedforward neural network are:

$$y_j^l = g_j\left(\sum_{i=1}^{m} y_i^{l-1} w_{ji}^l - b_j^l\right) \tag{7.1}$$

where y_j^l is the output of neuron j in layer l. g_j is the transfer function, typically a sigmoidal function. y_j^{l-1} is the output of neuron i in layer $l-1$ (containing m neurons) and w_{ji}^l the weight associated with that connection to neuron j. b_j^l is the internal threshold/bias of the neuron.

[2]The neuron bias offsets the sigmoidal transfer function; it is not to be confused with the inductive bias of the learning process.

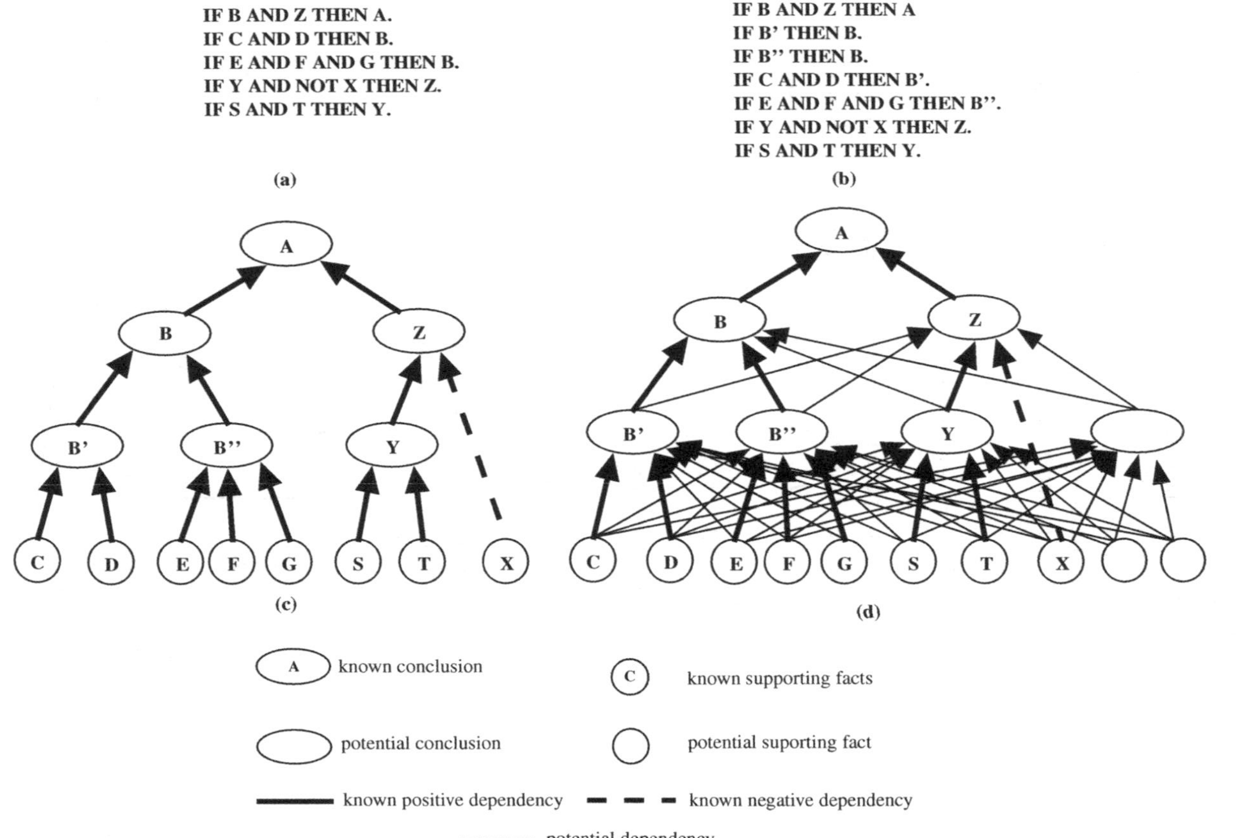

Fig. 7.1 Construction of KBANNs: **(a)** Original knowledge base, **(b)** rewritten knowledge base, **(c)** network constructed from rewritten knowledge base, **(d)** network augmented with additional neurons and weights

7.2 Inductive Bias

The KBANN prior knowledge encoding methodology introduces an inductive bias, H, that can be adjusted for better performance. The authors of the KBANN methodology chose the value of H by experimentation. A heuristic measure for determining H that takes the prior domain knowledge, the training data, the network architecture, and the learning method into account has been developed (Omlin et al., 2003 and Snyders et al., 2000). The heuristic uses gradient information of the error surface and its sensitivity to H, thus, $\partial E/\partial H$, to find a good encoding strength for the inductive bias H. The error function is:

$$E(H) = \frac{1}{2}(d_j - y_j{}^l(H))^2 \tag{7.2}$$

where d_j is the desired network output and $y_j^l(H)$ is the actual network output for a specific input pattern[3] p, where l is equal to the output layer. Notice that $y_j^l(H)$ depends on the particular choice[4] of H. Then, the derivative $\partial E/\partial H$ is given by

$$\frac{\partial E}{\partial H} = -(d_j - y_j^l)\frac{\partial y_j^l}{\partial H} = ErrSens_j^p \tag{7.3}$$

where l is equal to the output layer. We can compute $\partial y_j^l/\partial H$ recursively as follows:

$$\frac{\partial y_j^l}{\partial H} = g'_j \sum_{i=1}^{m} \left(\frac{\partial w_{ji}^l}{\partial H} y_i^{l-1} + w_{ji}^l \frac{\partial y_i^{l-1}}{\partial H} \right) \tag{7.4}$$

where w_{ji}^l connects neuron i, in layer $l-1$, with neuron j in the next hidden layer, l. g'_j is the derivative of the sigmoidal transfer function for neuron j. The derivative can $\partial w_{ji}^l/\partial H$ easily be calculated by

$$\frac{\partial w_{ji}^l}{\partial H} = \begin{cases} +1 \text{ if } w_{ji}^l = +H \\ -1 \text{ if } w_{ji}^l = -H \\ 0 \quad \text{otherwise} \end{cases} \tag{7.5}$$

The same equations apply for the neuron biases.

An overall sensitivity measure for the errors of all the output neurons and patterns is calculated from Eq. (7.3) as

[3]Normalization of the error function according to the number of patterns is necessary for a comparable value.

[4]This equation is for networks with a single output; the generalization to networks with multiple outputs is straightforward.

$$Sens_{EH} = \frac{\sum_{p=1}^{P} \sum_{j=1}^{J} ErrSens_j^P}{PJ} \tag{7.6}$$

Thus, varying H initially before training and calculating the sensitivity as given in Eq. (7.6) above, the heuristic suggests choosing the strength of the inductive bias H where the error surface in the direction of the prior knowledge, the direction of H, is the steepest:arg max $\left|\frac{\partial E(h)}{\partial H}\right|$, where $h \in \Phi$ and $\Phi = \left\{ H \middle| \frac{\partial^2 E(H)}{\partial H^2} = 0 \right\}$. This will prevent the network from traversing flat error surfaces during the initial training phase and suggests a good value for H other than through trial and error. After the best choice for H has been determined according to the heuristic, the domain knowledge is encoded using that value and then adapted through normal gradient descent techniques (e.g., the backpropagation learning algorithm).

7.3 Sensitivity Analysis and Measures

Although good empirical results (Omlin et al., 2003) have been achieved using the heuristic in the previous section, further analysis of the heuristic method is necessary for improved insight into its effectiveness.

7.3.1 Output-Pattern Sensitivity

One analysis would be to see how effective the heuristic is at encoding the prior domain knowledge such that the network will gain the most from the data presented to it during training. This could be determined through measuring the sensitivity of the output of the network with respect to the input training patterns. In Engelbrecht et al. (1995) and Zurada et al. (1994) methods for determining this sensitivity were developed.

The sensitivity, $Sens_{jz}^{p}$, of output neuron j with respect to input neuron z for a specific pattern p is defined as (Engelbrecht et al. 1995 and Zurada et al. 1994)

$$Sens_{jz}^{p} = \frac{\partial y_j^l}{\partial y_z^0} \tag{7.7}$$

where y_j^l is the output of neuron j in the output layer l and y_z^0 is the input of input neuron z.

We can compute $\partial y_j^l / \partial y_z^0$ recursively as follows:

$$\frac{\partial y_j^l}{\partial y_z^0} = g'_j \sum_{j=1}^{m} \left(w_{ji}^l \frac{\partial y_i^{l-1}}{\partial y_z^0} \right) \tag{7.8}$$

where w^l_{ji} connects neuron i, in layer $l-1$, with neuron j in the next hidden layer l. g'_j is the derivative of the sigmoidal transfer function for neuron j.

The recursive definition $\partial y^l_j / \partial y^0_z$ reduces in the following way

$$\frac{\partial y^l_j}{\partial y^0_z} = \begin{cases} 1 \ \textit{if}\ l = 0\ \textit{and}\ j = z \\ 0 \ \textit{if}\ l = 0\ \textit{and}\ j \neq z \end{cases} \tag{7.9}$$

Eq. (7.7) defines a sensitivity measure for a specific output-input neuron pair and for a specific pattern. We need to evaluate the sensitivity of the entire network, i.e., for all input patterns and all combinations of output-input neuron combinations. Engelbrecht et al. (1995) also define different metrics for combining the sensitivities given in Eq. (7.7). We chose the *absolute value average sensitivity* metric. The following equation calculates a normalized overall sensitivity value using that metric for summing all the sensitivities for different patterns and output-input neuron combinations:

$$Sens_{OP} = \frac{\sum_{p=1}^{P}\sum_{j=1}^{J}\sum_{z=1}^{Z}\left|Sens^P_{jz}\right|}{PJZ} \tag{7.10}$$

7.3.2 Output-Weight Sensitivity

The sensitivity of output neuron j with respect to a specific weight w_{xy} in the network for a given pattern p is given by

$$Sens^P_{jxy} = \frac{\partial y^l_j}{\partial w^u_{xy}} = g'_j \sum_{i=1}^{m}\left(\frac{\partial y^{l-1}_i}{\partial w^u_{xy}} w^l_{ji} + y^{l-1}_i \frac{\partial w^l_{ji}}{\partial w^u_{xy}}\right) \tag{7.11}$$

Eq. (7.11) reduces to the following

$$\frac{\partial y^l_j}{\partial w^u_{xy}} = \begin{cases} g'_j y^{l-1}_i & \textit{if}\ l = u,\ j = x,\ \textit{and}\ i = y \\ g'_j \sum_{i=1}^{m} \frac{\partial y^{l-1}_i}{\partial w^u_{xy}} w^l_{ji} & \textit{otherwise} \end{cases} \tag{7.12}$$

The same equations also apply to neuron biases. For an overall sensitivity measure, the *absolute value average* is taken of all the sensitivities for all the output neurons, weights/biases, and patterns:

$$Sens_{OW} = \frac{\sum_{p=1}^{P}\sum_{j=1}^{J}\sum_{x=1}^{X}\sum_{y=1}^{Y}\left|Sens^P_{jxy}\right|}{PJZ} \tag{7.13}$$

7.3.3 Output-H Sensitivity

The sensitivity of the network output with respect to the *inductive bias H* for a specific pattern can be measured with Eq. (7.14):

$$Sens_j^p = \frac{\partial y_j^l}{\partial H} \tag{7.14}$$

An overall sensitivity measure for all the output neurons and patterns is calculated by

$$Sens_{OH} = \frac{\sum_{p=1}^{P}\sum_{j=1}^{J} Sens_j^P}{PJ} \tag{7.15}$$

7.3.4 Euclidean Distance

Determining the *inductive bias H* using the heuristic method described supposedly suggests a good starting point for training the network. The premise of the heuristic is that good local minima in the error surface would be found in steep ravines rather than in shallow valleys. Thus, a pertinent question to ask is how to analyze whether the method actually suggests a good starting position in weight space for training the network.

One measure for training effort is the number of epochs needed. Others are the distances traversed in weight space due to weight updates, i.e., from each successive weight vector to the next, and the distance between the initial position in weight space before training and the final position after training. We therefore also computed the *total* (incrementally summed) Euclidean distances followed through the whole training procedure by adding the distances together after each weight update, and computed the direct (denoted *netto*) Euclidean distance between the initial starting position (represented by the prior knowledge encoding) and the final trained position of the network's weights.

7.4 Promoter Recognition

We applied our analysis of the heuristic for choosing the *inductive bias H* to the published problem (Towell et al. 1994) of identifying prokaryotic promoter sites in sequenced DNA (see also Snyders et al., 2000).

7.4.1 Data and Initial Domain Theory

The data for the recognition of promoters were used from the machine learning repository of the University of California (Murphy et al. 1994). The data set consisted of 53 positive and 53 negative examples.

The rules for the promoter recognition task are listed in Table 8.2. For our purposes, we represent DNA as a linear sequence of nucleotides from the set {A, G, T, C}. The rules use a notation to specify where a sequence of DNA is likely to occur relative to a reference point. This reference point occurs seven nucleotides to the left of the end of the DNA sequence. Thus, the notation @40 'AT-C' means that an 'A' must appear 40 nucleotides to the left of the reference point, a 'T' must appear 39 nucleotides to the left of the reference point, etc. The '-' indicates that any nucleotide will suffice.

According to the rule set, there are two sites at which the *RNA polymerase* binds to the DNA, minus-10 and minus-35[5]. The conformation rule attempts to simulate the three-dimensional structure of DNA and to make sure that the minus-10 and minus-35 sites are spatially aligned.

The initial domain theory in Table 7.1 was encoded using the KBANN architecture. The structure of the network, before the addition of low-weighted random-initialized weights, is shown in Fig. 7.2. For a sequence location, four input units were programmed to represent the set {A, G, T, C}.

Table 7.1 Knowledge base for promoter recognition: The rules, in Prolog notation, specify where a sequence of DNA is likely to occur relative to a reference point

```
promoter        :- contact, conformation
contact         :- minus-35, minus -10.
minus-35        :- @-37 'CTTGAC-'.
minus-35        :- @-37 '-TTGACA'.
minus-35        :- @-37 '-TTG-ACA'.
minus-35        :- @-37 '-TTGAC-'.
minus-10        :- @-14 '--TA---T'.
minus-10        :- @-14 '-TA-A-T'.
minus-10        :- @-14 '-TATAAT-'.
minus-10        :- @-14 'TATAAT--T'.
conformation    :- @-45 '-AA--A'.
conformation    :- @-45 '-A----A', @-28 -'T---T-AA--T-'
                   ,@-04 'T'.
conformation    :- @-49 '-A----T', @-27 'T----A--T-TG-'
                   ,@-01 'A'.
conformation    :- @-47 'CAAT-TT-AC', @-22 'G---T-C' ,@-08
                   'GCGCC-CC'.
```

[5]These two rules are named according to their position from the reference point.

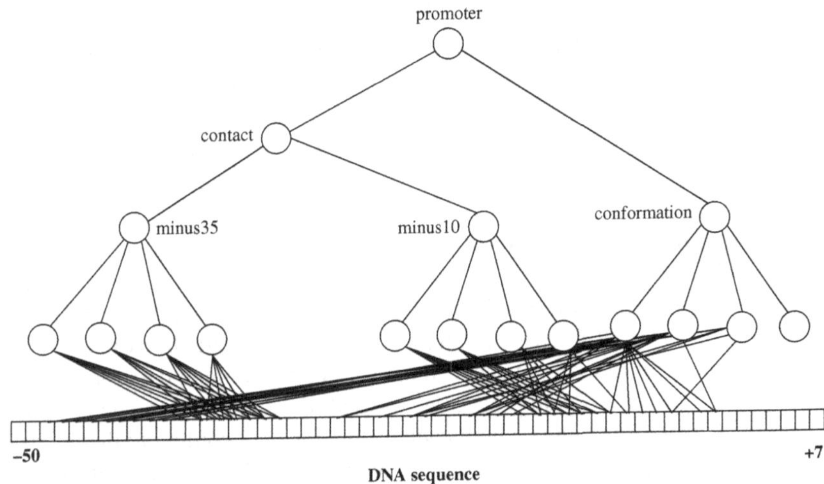

Fig. 7.2 KBANN for promoter recognition: The structure of the knowledge-based neural network derived from the rules in Table 7.1. The neurons corresponding to minus 35, minus 10, and conformation are disjunctive; all the other neurons correspond to conjunctive rules. Random-initialized, low-weighted links are not shown

7.4.2 Experimental Methodology

We used KBANNs with neurons with sigmoidal transfer functions and the standard quadratic error function E (refer to Eq. 7.2). A network correctly classified an example, during training, if its output was within 0.25 of the desired output, and for testing, within 0.5 of the desired output. We chose the learning rate 0.1 and the momentum 0.1, and trained all networks until all the patterns in the training set were correctly classified.

We performed a ten-fold cross-validation on the data. Each fold contained 96 of the examples from the data set (except the last fold, which had 90 examples); the remaining examples were used for testing. We encoded networks with the domain knowledge, varying H from 0.1 to 7.0 in increments of 0.1. For each of these encoded but untrained networks we measured on the training set:

- The error ratio
- The sensitivity of the error with respect to H (heuristic for determining inductive bias) according to Eq. (7.3),
- The network output sensitivity with respect to the input patterns according to Eq. (7.10),
- The network output sensitivity with respect to all the weights in the network according to Eq. (7.13),
- And the network output sensitivity with respect to H according to Eq. (7.15).

We then trained the networks and measured:

- The training time in epochs,
- The *total* Euclidean distance followed through the training process,
- The *netto* Euclidean distance between the initial encoded network and the final trained network,
- And the generalization performance on the test set

7.5 Discussion and Conclusion

Figures 7.3 to 7.6 show the results of the experiments for one fold of the ten, which are typical for all folds. These figures also show the graphs of the function $\partial E/\partial H$ and the error ratio of the encoded network on the training data for comparison purposes. The various graphs have been scaled so that the data can be compared easily. Fig. 7.3 shows that the recommended value for H, according to the heuristic of Sect. 7.3, is about 1.8 at the peak of the function $\partial E/\partial H$. At this point the error of the encoded network on the training data has already reached its minimum value, and one can notice the sharp drop in the error down from 100% the moment that the encoded knowledge comes into play. At lower values for H the knowledge is simply not encoded strongly enough. The error on the test set for this fold varies up and down for $0.1 \leq H \leq 1$, but is relatively low and constant for values of H ranging from about 1 to 2.2, and only starts to increase markedly from about $H \geq 3$. The number of epochs needed to reach 100% correctness on the training set increases significantly for values of H greater than 4.

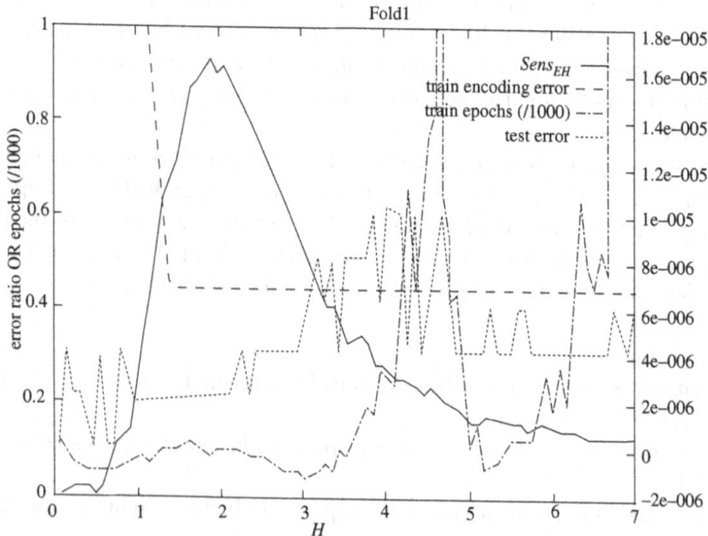

Fig. 7.3 Typical result from Fold 1: For networks with different values of the inductive bias H, the encoding error, training time (epochs), and error on the test set are plotted. It also plots the function $\partial E/\partial H$ as a function of the inductive bias strength H

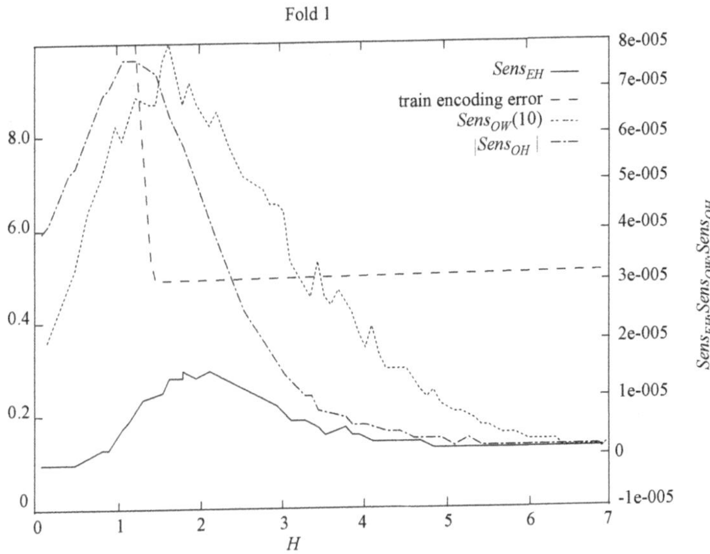

Fig. 7.4 Typical result from Fold 1: For networks with different values of the inductive bias H, the encoding error, output-weights sensitivity, output-H sensitivity and the function $Sens_{EH}$ are shown

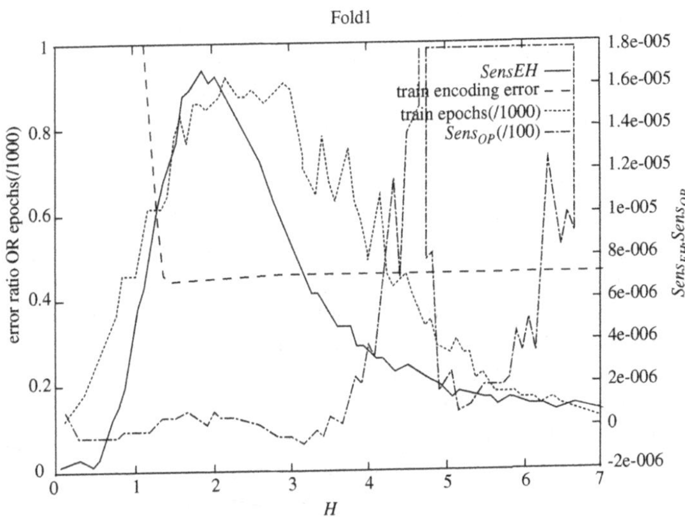

Fig. 7.5 Typical result from Fold 1: For networks with different values of the inductive bias H, the encoding error, training time, the output-pattern sensitivity, and $Sens_{EH}$ are plotted

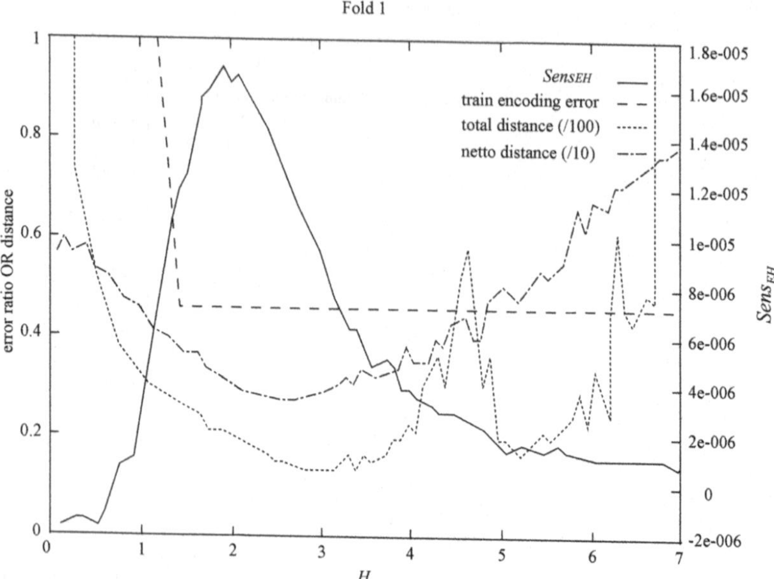

Fig. 7.6 Typical result from Fold 1: For networks with different values of the inductive bias H, the encoding error, *total* Euclidean distance, *netto* Euclidean distance, and $Sens_{EH}$ are plotted

Figure 7.4 plots the network output sensitivity with respect to all the randomly set weights in the network (i.e., the weights preset to H are excluded) according to Eq. (7.13), and the network output sensitivity with respect to H according to Eq. (7.15). When the output of a sigmoidal transfer function is close to 0 or to 1 its derivatives become much smaller than the maximum value of 0.5. In this case it is much more difficult to train a network with backpropagation (due to its gradient descent weight update method) when the transfer functions approach their asymptotic values. For subsequent training of an encoded network, the error can be measured on average using the sensitivity measures depicted in the graphs. The higher parts of both graphs lie approximately in $0.8 \leq H \leq 2.4$, showing that the suggested heuristic value of H leaves the network trainable, but not with saturated transfer functions on average. Notice that the graph of Eq. (7.15) diminishes quicker than that of Eq. (7.13). It is desirable to choose a value for H where it is less sensitive to training than the values of the randomly initialized weights, in order not to destroy the prior knowledge encoded into the network.

Figure 7.5 plots the network output sensitivity with respect to the input patterns according to Eq. (7.10), and includes the number of training epochs again. The most sensitive range for the input patterns is about $1.3 \leq H \leq 3.2$, showing that the input patterns can still successfully be used for further refinement of the network, and that training does not require an excessive amount of epochs.

Figure 7.6 plots the *total* Euclidean distance followed through the training process and the *netto* Euclidean distance between the initial encoded network's and the final trained network's positions in weight space as additional measures of the training effort. The most direct path to the final weight vector is reached for H set to about 2.6, which is well positioned within the most sensitive area pointed to by Eq. (7.10) in Fig. 7.6.

To conclude, alternative good values for setting H to encode the prior knowledge correspond to the areas of the sensitivity measures around their peaks. Notably, from the graph of Eq. (7.10) in Fig. 7.6, good values for H are also in the range $1.5 \leq H \leq 3$, illustrating the usefulness of sensitivity analysis metrics for understanding the encoding of prior knowledge into a neural network.

Chapter 8
Applications

8.1 Input Dimension Reduction

The curse of dimensionality is always problematic in pattern classification problems. In this chapter, we provide a brief comparison of the major methodologies for reducing input dimensionality and summarize them in three categories: correlation among features, transformation and neural network sensitivity analysis. Furthermore, we propose a novel method for reducing input dimensionality that uses a stochastic RBFNN sensitivity measure. The experimental results are promising for our method of reducing input dimensionality.

There are typically two ways to reduce input dimensionality, namely, the correlation approach and the transformation approach (Ng and Yeung, 2002). The former reduces the number of highly correlated features, whereas the latter maps the original input space into an orthogonalized space, in which the principal dimensions are selected.

In the correlation approach, the covariance of each pair of features is evaluated to create a covariance matrix. The higher the covariance of two features, the higher the dependence between them. One of the two dependent features is eliminated to leave one representative feature in the subset of input features. The subset of remaining input features contains those features that have a low correlation among them. Battiti (1994) applied mutual information to supervised reduction of the input dimensionality, where the mutual information indicates the dependencies and the information gain within the input features. Kwok and Choi (2002) improved on Battiti's method by combining the mutual information and Taguchi methods to reduce the input dimensionality. In all of these methods, each input feature, of all sample points, is evaluated one by one. This means that all methods in this category require a greedy search heuristic, which is prohibited by the high computational cost.

In the transformation approach, the set of original input features is transformed into, or projected onto, a lower-dimensional space to reduce their dimensionality. Principal component analysis (PCA) (Jolliffe, 1986) and independent component analysis (ICA) (Hyvärinen and Oja, 2000) are two well-known tools for transforming the existing input features into a new lower-dimension feature space. In PCA, the input feature space is transformed into a lower-dimensional feature space using

D.S. Yeung et al., *Sensitivity Analysis for Neural Networks*, Natural Computing Series,
DOI 10.1007/978-3-642-02532-7_8,

the largest eigenvectors of the correlation matrix. In ICA, the original input space is transformed into an independent feature space with a dimension that is independent of the other dimensions. A wavelet transform, or simply a Fourier transform, transforms the input features into the frequency domain. The frequency component in the input features can be separated and frequency bands with less information can be eliminated to reduce the dimensions of the input.

The problem with the correlation approaches is their high computational complexity. The complexity is $O(n^2)$, where n is the number of samples, because of the greedy search heuristics. The transformation approaches that use transformations are relatively fast and easy to use, but the description of the output is no longer provided by the original input features. Data mining does not favor this kind of approach, although it is commonly used in signal, image and audio processing. Furthermore, the transformation still requires the full set of the original data, which does not save time and cost in data collection and storage.

To address the problems caused by the correlation and transformation approaches, some researchers have employed sensitivity analysis to input dimensionality reduction (Zurada, Malikowski and Usui, 1997).

8.1.1 Sensitivity Matrix

It can be easily noticed that the entries of the Jacobian matrix defined in Equation (1.3) can be considered as sensitivity coefficients. Especially, sensitivity of an ouput o_k with respect to its input x_i is

$$S_{ki} = \frac{\partial o_k}{\partial x_i} \tag{8.1}$$

By using standard notation of an error backpropagation approach, the derivative of Eq. (8.1) can be readily expressed in terms of network weights as follows:

$$\frac{\partial o_k}{\partial x_i} = o'_k \sum_{j=1}^{J-1} w_{kj} \frac{\partial y_j}{\partial x_i} \tag{8.2}$$

where y_j denotes the output of the jth neuron of the hidden layer, and o'_k is the value of the derivative of the activation function $o = f(\text{net})$ taken at the k th output neuron. This further yields

$$\frac{\partial o_k}{\partial x_i} = o'_k \sum_{j=1}^{J-1} w_{kj} y'_j v_{ji} \tag{8.3}$$

8.1.2 Criteria for Pruning Inputs

Inspection of the mean square average (MSA) sensitivity matrix S_{avg} allows us to determine which inputs affect outputs least. A small value of $S_{\mathrm{ki,\,avg}}$ in comparison to others means that for the particular kth output of the network, the ith input does not significantly contribute per average to output k, and therefore could be possibly disregarded. This reasoning and results of experiments allow us to formulate the following practical rules: The sensitivity matrices for a trained neural network can be evaluated for both training and testing data sets; the norms of MSA sensitivity matrix columns can be used for ranking inputs according to their significance and for reducing the size of the network accordingly through pruning less relevant inputs.

When one or more of the inputs have relatively small sensitivity in comparison to others, the dimension of neural network can be reduced by removing them, and a smaller-sized neural network can be successfully retrained in most cases. Suppose two inputs provide important data for a neural network; one of them has much larger relative change than the other. In such a case, the sensitivity of the second output would be much larger than the first due to the necessity of an additional amplification of the input by network weights. In the extreme, the first of those two inputs may even be selected for pruning. To prevent such cases, the following formulas allow us to scale inputs into the range of $[-1, 1]$.

$$\hat{x}_i^{(m)} = \frac{x_i^{(m)} - \left(\left(\max\limits_{n=1,\cdots,N}\{x_i^{(n)}\} + \min\limits_{n=1,\cdots,N}\{x_i^{(n)}\}\right)/2\right)}{\left(\max\limits_{n=1,\cdots,N}\{x_i^{(n)}\} - \min\limits_{n=1,\cdots,N}\{x_i^{(n)}\}\right)} \tag{8.4}$$

$$\widehat{o}_k{}^{(m)} = \frac{o_i{}^{(m)} - \left(\left(\max\limits_{n=1,\cdots,N}\{o_i{}^{(n)}\} + \min\limits_{n=1,\cdots,N}\{o_i{}^{(n)}\}\right)/2\right)}{\left(\max\limits_{n=1,\cdots,N}\{o_k{}^{(n)}\} - \min\limits_{n=1,\cdots,N}\{o_k{}^{(n)}\}\right)} \tag{8.5}$$

where ^ denotes the normalized variable, or parameter.

8.2 Network Optimization

In training neural network, training data are usually finite and nonuniformly sampled, so the problem is consequently ill-posed. Conversion to a well-posed problem is typically achieved with some form of capacity control, which aims to balance the fitting of the data with constraints on the model flexibility, producing a robust model that generalizes successfully (Gao, Harris and Gunn, 2001). In practice, such an optimization is accomplished by searching for the simplest network architecture under the well-recognized Occam's Razor hypothesis: "*plurality should not be posited without necessity*," or in other words, the simpler a solution is, the more reasonable it is. This section aims to apply sensitivity analysis to find the optimal network architecture to achieve the highest possible generalization capability and the lowest possible system complexity.

Engelbrecht (2001) proposed a sensitivity-based pruning technique. The relevance of parameters is quantified based on parameter sensitivity information, and the irrelevant parameters are subsequently removed. A variance nullity measure is computed for each parameter, based on ideas borrowed from the nonconvergent tests of Finnoff, Hergert and Zimmermann (1993). The basic idea of the variance nullity measure is to test whether the variance in parameter sensitivity for the different patterns is significantly different from 0. If the variance in parameter sensitivities is not significantly different from 0 and the average sensitivity is small, the corresponding parameter has little or no effect on the output of the neural network over all patterns considered. A hypothesis testing step then uses these variance nullity measures to statistically test if a parameter should be pruned.

What needs to be done is to test whether the expected value of the sensitivity of a parameter over all patterns is equal to 0. The expectation can be written as

$$\left\langle S^2_{o\theta,ki} \right\rangle = \langle S_{o\theta,ki} \rangle^2 + \mathrm{var}\left(S_{o\theta,ki}\right) \tag{8.6}$$

where $S_{o\theta,ki}$ refers to the sensitivity of output o_k to changes in parameter θ_i for a single pattern p.

Define the statistical nullity in the parameter sensitivity variance of a neural network parameter θ_i over patterns $p = 1, 2 \ldots P$ as

$$\gamma_{\theta_i} = \frac{(P-1)\sigma^2_{\theta_i}}{\sigma^2_0} \tag{8.7}$$

where $\sigma^2_{\theta_i}$ is the variance of the sensitivity of the network output to perturbations in parameter θ_i and σ^2_0 is a value close to 0 (the characteristics of this value are explained below).

The *variance* in parameter sensitivity $\sigma^2_{\theta_i}$ is computed as

$$\sigma^2_{\theta_i} = \frac{\sum_{p=1}^{P} \left(N^{(p)}_{\theta_i} - \bar{N}_{\theta_i}\right)^2}{P-1} \tag{8.8}$$

where

$$N^{(p)}_{\theta_i} = \frac{\sum_{k=1}^{K} S^{(p)}_{o\theta,ki}}{K} \tag{8.9}$$

and $\bar{N}_{\theta_i}$ is the *average* parameter sensitivity

$$\bar{N}_{\theta_i} = \frac{\sum_{p=1}^{P} N^{(p)}_{\theta_i}}{P} \tag{8.10}$$

Using the fact that under the null hypothesis the variance nullity measure has an $\chi^2(P-1)$ distribution in the case of P patterns, the critical value γ_c is obtained from χ^2 distribution tables

$$\gamma_c = \chi^2_{v;1-\alpha} \tag{8.11}$$

where $v = P - 1$ is the number of degrees of freedom and α is the level of significance. The complete pruning algorithm is given in Fig. 8.1.

The variance nullity algorithm starts pruning the hidden layer first, then the input layer. Pruning consists of removing the irrelevant hidden units or input units and all the weights leading from those units to the next layer. Calculation of the variance nullity measures can be done on the training, validation, or test sets. During

STEP 1 Initialize the NN architecture and learning parameters.

STEP 2 Repeat

(a) train the NN until a pruning indicator is triggered;

(b) let σ_0^2 = 0.0001;

(c) for each θ_i,

(i) for each p=1,2, …, P, calculate $N_{\theta_i}^{(p)}$ using (8.9);

(ii) calculate the average $\overline{N}_{\theta_i}$ using (8.10);

(iii) calculate the variance in parameter sensitivity using $\sigma_{\theta_i}^2$ from (8.8);

(iv) calculate the test variable γ_{θ_i} using (8.7).

(d) apply the pruning heuristic:

(i) arrange γ_{θ_i} in increasing order;

(ii) find γ_c using (8.11);

(iii) for each θ_i, if $\gamma_{\theta_i} \le \gamma_c$, then prune θ_i ;

(iv) if $\gamma_{\theta_i} > \gamma_c$ for all θ_i,

then let $\sigma_0^2 = \sigma_0^2 \times 10$ and goto step 2(c)(iv).

(v) Until no θ_i is pruned, or the reduced network is not accepted due to an unacceptable deterioration in generalization performance.

STEP 3 Train the final pruned NN architecture.

Fig. 8.1 Pruning algorithm based on variance nullity

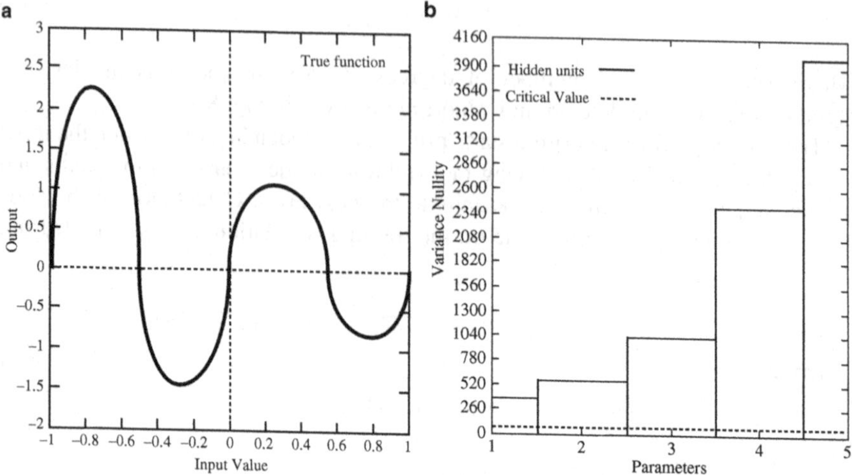

Fig. 8.2 Experimental results on the function $f(x) = \sin(2\pi x)e^{-x} + \varsigma$. (a) Function curve. (b) Hidden unit variance nullity at the end of pruning (Engelbrecht, 2001)

training, the error on a validation set is monitored to detect when overfitting starts, at which point pruning is initiated. When a network is pruned, the pruning model starts retraining of the reduced model on new initial random weights. The values of pruned weights are not redistributed over the remaining weights. The pruning stopping criterion is encapsulated in the pruning heuristic: If no more parameters can be identified for pruning, or if a reduced model is not accepted, the pruning process terminates.

Figure 8.2a shows the function $f(x) = \sin(2\pi x)e^{-x} + \varsigma$, on which the hypothesis testing is applied. Fig. 8.2b shows the experimental results.

From Fig. 8.2b, it can be seen that the number of hidden units sufficient to learn a given target function is equal to the number of turning points of that function plus 1 (if sigmoid activation functions are used). An activation function is fitted for each inflection point. The minimum number of hidden units for the function is therefore five (Fig. 8.2b). The next pruning step (using a larger value for verification) resulted in the removal of all the hidden units, for which the network could not learn the function. The results illustrate that the statistical pruning heuristic correctly removed the redundant hidden units.

8.3 Selective Learning

Active learning or sequential learning is defined as any form of learning in which the learning algorithm has some control over a certain part of the input space (Cohn, Atlas and Ladner, 1994). An active learning strategy allows the learner to dynamically select training examples from a candidate training set. By adding this functionality to a neural network, we change the network from a passive learner to an active learner. There are two approaches to active learning, namely, incremental learning and selective learning. The former starts training on an initial subset of a candidate training set, from which the trained examples in selected subsets are removed. In contrast, selective learning selects a new training subset at each selection interval from the original candidate set, using the current knowledge of the network about the problem domain.

This section introduces the sensitivity analysis selective learning algorithm (SALSA), which uses sensitivity analysis to select patterns in the region of a decision boundary. First-order derivatives of the output units with respect to the input units are used to determine how close a pattern lies to a decision boundary. Patterns which lie closest to the decision boundaries, which are the most *informative patterns*, are selected for training (Engelbrecht, 2001).

First of all, let us define the informativeness of a pattern as the sensitivity of the neural network output vector to small perturbations in the input vector:

$$\Phi^{(p)} = \left\| S_{oz}^{(p)} \right\| \tag{8.12}$$

where $S_{oz}^{(p)}$ is the output-input layer sensitivity matrix. Assuming sigmoid activation functions in the hidden and output layers, each element $S_{oz,ki}^{(p)}$ of the sensitivity matrix is computed using

$$S_{oz,ki}^{(p)} = \left(1 - o_k^{(p)}\right) o_k^{(p)} \sum_{j=1}^{J} w_{kj} \left(1 - y_j^{(p)}\right) y_j^{(p)} v_{ji} \tag{8.13}$$

where w_{kj} is the weight between output unit O_k and hidden unit y_j, v_{ji} is the weight between hidden unit y_j and input unit z_i, $o_k^{(p)}$ is the activation function of output o_k, $y_j^{(p)}$ is the activation of hidden unit y_j, and J is the total number of hidden units.

A pattern is considered informative if any one or more of the output units are sensitive to small perturbations in the input vector. The larger the value of $\Phi^{(p)}$, the more informative the pattern p.

Next, according to the calculation of informativeness, let us see how the patterns are selected for training. The objective of a neural network classifier is to construct optimal decision boundaries over input space. Active learning algorithms which sample from a region around decision boundaries have been shown to refine boundaries, resulting in improved generalization performance. For each pattern p, the sensitivity of each output unit o_k to small perturbations in each input unit z_i is $S_{oz,ki}^{(p)}$. Patterns with high $S_{oz,ki}^{(p)}$ values lie closest to decision boundaries. The sensitivity analysis selective algorithm makes use of this fact to assign to each pattern a "measure of closeness" to decision boundaries. Patterns closest to decision boundaries are the most informative. Selecting the most informative patterns therefore results in training only on patterns close to boundaries (Engelbrecht, 1999b).

8.4 Hardware Robustness

As mentioned in Chap. 2, Stevenson (1990) introduced the hypersphere model as a geometrical approach to sensitivity analysis. To complete the computation for a specific case, let us consider a single-step activation function. Referring to Fig. 2.3 of the hypersphere approximation, there are only one hypersphere and two lunes in this case. Since the lunes are centered at the origin, the ratio of the area of the lunes to the area of the zones is equal to the ratio of the rotation angle θ to π. By replacing θ with its expected value, we obtain

$$P_w(\delta W) = \frac{E(\theta)}{\pi} = \frac{1}{\pi}\int_0^{\pi} arctg\left(\frac{\delta W \cdot \sin\phi}{1 + \delta W \cdot \cos\phi}\right) \cdot p(\phi)d\phi \tag{8.14}$$

where $p(\phi)$ is the probability density function of the random variable, defined as in Winter (1989):

$$p(\phi) = \frac{K_n}{K_{n+1}} \cdot \sin^{n-1}\phi \tag{8.15}$$

where $K_n = 2\pi^{n/2}/\Gamma(n/2)$, and $\Gamma(\bullet)$ is the Gamma function. Integration of Eq. (8.14) leads to (Alippi, Piuri and Sami, 1994):

$$P_w(\delta W) = \frac{1}{2\pi}\left[\frac{\pi}{2} + 2 \cdot arctg\left(\frac{\delta W - 1}{\delta W + 1}\right)\right] \tag{8.16}$$

for any value of δW. The probability density function is obtained for each neuron in the output layer. Integration of each such function over the whole range of possible values of the output error ratio provides the probability of observing an error at the output of the corresponding neuron. To have a first approximation of the network's sensitivity, the maximum value of all probabilities can be considered; in turn, this provides a lower bound for the error probability.

To verify the above methodology with hardware robustness, Alippi, Piuri and Sami (1995) train a simple three-layered network composed of 20 inputs, 15 neurons in the hidden layer, and one output neuron, to detect the presence of a ship in a radar image. A preliminary VLSI design has been carried out, adopting a 0.7 μm CMOS technology; eight bits have been chosen for weight representation and 13 bits for the adders (to avoid truncation problems). A defect distribution of 15 defects per square centimeter has been assumed as a reasonable parameter. As a consequence, the information summarized in Table 8.1 has been derived.

In Table 8.1, for different classes of defects, the total number of defects in each class is evaluated as related to the silicon area occupied by the digital devices, the relative occurrence probability of each defect class, the probability that an error (due to the defects) propagates to the network's output, and finally, the probability of

Table 8.1 A sample evaluation of the error probability in the presence of production-time defects

Defect Type	Defect Number	Defect Probability	Error Propagation Probability	Output Error Probability
Hidden neuron's memory	0.6670	0.2081	0.2478	0.0516
Output neuron's memory	0.0334	0.0105	0.5033	0.0168
Hidden neuron's multiplier	0.4104	0.1280	0.2480	0.0317
Output neuron's multiplier	0.0205	0.0064	0.5017	0.0032
Hidden neuron's adder	1.9755	0.6163	0.0267	0.0165
Output neuron's adder	0.0987	0.0307	0.1171	0.0036

observing an error in the network's output due to the defect distribution. (This probability is the product of the defect occurrence probability and the error propagation probability.) From Table 8.1, we can see that the identification of the weight memory of the hidden layer is the most critical section in the design (Alippi, Piuri and Sami, 1995).

8.5 Measure of Nonlinearity

In this application, the variance of the sensitivity of neural network output to input parameter perturbations is used as a measure of the nonlinearity of the data set. This measure of nonlinearity is then used to show that the higher the variance of noise injected to output values, the more linearized the problem.

Multilevel models are an important tool for exploring nonlinearity in response in hierarchically structured data (Goldstein, 1995). They enable, among other things, the study of effects in a population of individuals, taking into account variations between the individuals' responses, for instance, in a study design, with repeated measurements for each individual. Since they are based on investigating the distribution of model parameters over the population of individuals, they are restricted to parametric models. In practice, most multilevel studies use linear effect models, possibly relative to log-transformed values of the variables, often despite a lack of evidence of linearity of effects. Considering this, a neural network based multilevel approach seems an interesting nonlinear alternative to commonly used analyses. Unfortunately, many real-world data gathering systems are very noisy. For example, a measurement device may not correctly report the value of interest, or the measured value may not completely agree with the physical quantity it is supposed to describe. As an example of the latter, consider a study in which ambient ozone measurements are used as the estimates of an individual's personal exposure to this pollutant. An individual's true personal exposure, however, is determined by many factors that are not taken into account in this setup, and using such a crude estimate of individual exposure introduces strong error (Lebret, 1990).

Lamers, Kok and Lebret (1998) introduce perturbation analysis to solve the problem. They hypothesize that high levels of measurement noise added to a nonlinear system may lead to *linearization* of that system, meaning that nonlinear effects are obscured by noise, and appear to be more linear. Using the proposed nonlinearity study, we are able to test this hypothesis in simulations based on a nonlinear model of an epidemiological system to which noise is added. In the model, the response of each individual in a population is modeled by a separate nonlinear feedforward network. A measure of nonlinearity in response is assigned to each such network, and the distribution of this parameter over the studied population is examined.

Let us assume that only one input variable is used in the model, and discard the variable index v for notational convenience wherever possible. Although the values of other input variables may affect the (non)linearity in response to the studied variable, we can restrict our attention to univariate models without loss of generality. For every individual we compute the gradient

$$g^i_\mu = \frac{\partial O^i}{\partial v}(\mu) \tag{8.17}$$

for each observation μ, and derive its sample variance over all observations:

$$\gamma^i = \mathrm{var}_\mu(g^i_\mu) = \sum_\mu \left(g^i_\mu - \overline{g^i}\right)^2 \frac{n_i}{n_i - 1} \tag{8.18}$$

where O^i is the output of network i, and n_i is the number of observations obtained from individual i.

The γ coefficient expresses a quantitative notion of linearity in the response of feedforward networks. It incorporates nonparametric nonlinear models into the lower level of a multilevel study. The method is independent of the study domain, and applies in all studies in which the data exhibit hierarchical structure.

8.6 Parameter Tuning for Neocognitron

Fukushima's Neocognitron is well known for its performance in visual pattern recognition (Fukushima 1982, 1987, 1988). The Neocognitron divides the visual patterns into many sub-patterns (features) and distributes them into different feature planes which are linked in a hierarchical manner. The Neocognitron consists of two types of neurons, i.e., S-cells, which undertake feature extraction, and C-cells, which take charge of the toleration of the deformation. There are two important parameters for each S-cell, one is its receptive field, which specifies the number of its input interconnection, and the other is its selectivity, which is a threshold to control its response to features.

Many researchers, including Fukushima himself, have made efforts to improve the performance of the Neocognitron. In this chapter, sensitivity analysis will be applied to Neocognitron to find out how its parameters affect the performance, and then a genetic algorithm based technique will be employed to tune the parameters.

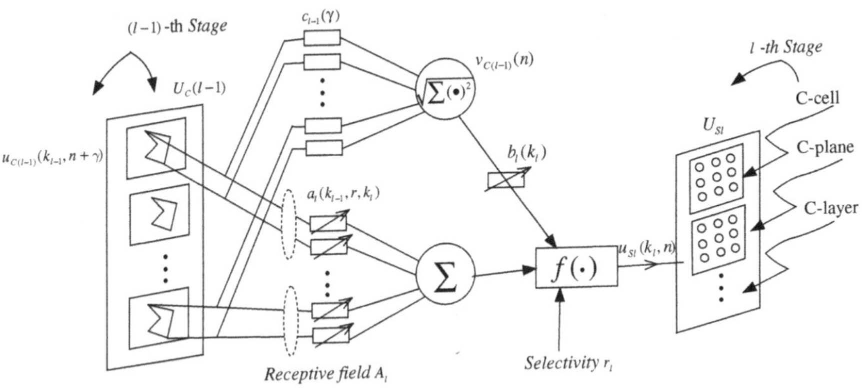

Fig. 8.3 Hierarchical relationship among cell, plane, layer and stage in the Neocognitron

The Neocognitron tries to model the visual system which can recognize an object even if the object shifts its position and is slightly deformed. The Neocognitron can learn from the stimulus patterns with shift invariance ability and can be generalized to recognize any unseen deformed version of these patterns. The Neocognitron consists of simple cells, called *S-cells*, and complex cells, called *C-cells*. In Fig. 8.3, S-cells are grouped together to form *S-planes*, which are combined to form an *S-layer*, denoted as *Us*. A similar relationship exists among C-cells, *C-planes*, and *C-layer(Uc)*. S-layer and C-layer are put together to form an intermediate stage of the network. In the input stage, the only layer involved is a two-dimensional plane U_O which accepts two-dimensional visual patterns directly. It simplifies knowledge representation for the input. In fact, the feature extraction step has been embedded in the connection scheme of the Neocognitron model. For character recognition, the output stage consists of the S-layer and the C-layer, where every C-plane has only one C-cell which represents a category of characters.

8.6.1 Receptive Field

In the Neocognitron, only the input interconnections to S-cells are variable and modifiable; the input interconnections to other cells are fixed and unmodifiable. Fukushima uses the notation u_{Sl} (k_l; **n**) to represent the output of an S-cell in the k_lth S-plane in the lth stage, and u_{Cl} (k_l; **n**) to represent the output of a C-cell in the k_lth C-plane in that stage, where **n**is the two-dimensional coordinates representing the position of the cell's *receptive fields*on the input layer.

From the biological perspective, the visual stimulus light passes through the eyes and hits the retina. The retina ganglion cell collects the signals from a group of photoreceptors which form the receptive field of the cell. The retina ganglion cell sends signals to lateral geniculate body (LGB) so that the signals are arranged correctly for transmission in the visual cortex. This process is simulated by an input layer in the Neocognitron. The size of the receptive field is set to be small in the first stage and to increase with the depth l.

8.6.2 Selectivity

In the Neocognitron, S-cells have inhibitory inputs with a shunting mechanism. The output of an S-cell of the k_l th S-plane in the lth stage is given by:

$$u_{Sl}(k_l,n) = r_l \cdot \varphi \left[\frac{1 + \sum\limits_{k_{l-1}=1}^{K_{l-1}} \sum\limits_{\gamma \in A_l} a_l(k_{l-1},\gamma,k_l) \cdot u_{C(l-1)}(k_{l-1},n+\gamma)}{1 + \frac{r_l}{1+r_l} \cdot b_l(k_l) \cdot v_{C(l-1)}(n)} - 1 \right] \quad (8.19)$$

where $\varphi[x] = \max(x,0)$. $a_l(k_{l-1},\gamma,k_l)$ and $b_l(k_l)$ represent the values of the excitatory and inhibitory variable interconnecting coefficients, respectively.

Parameter γ_l in Eq. (8.19), called selectivity, controls the intensity of the inhibition. The larger the value γ_l, the more selective the cell's response to its specific feature. The inhibitory cell V-cell, $v_{C(l-1)}(\mathbf{n})$, which is sending an inhibitory signal to cell $u_{Sl}(k_l;\mathbf{n})$, receives its input interconnections from the same cells as $u_{Sl}(k_l;\mathbf{n})$ does, and yields an output proportional to the weighted mean square root of its inputs:

$$v_{C(l-1)}(n) = \sqrt{\sum_{k_{l-1}}^{K_{l-1}} \sum_{\gamma \in A_l} c_{l-1}(\gamma) \cdot u^2_{C(l-1)}(k_{l-1},n+\gamma)} \quad (8.20)$$

The values of fixed interconnections $c_{l-1}(\gamma)$ are determined so as to decrease monotonically with respect to the value of $|\gamma|$ but to satisfy $\sum\limits_{k_{l-1}}^{K_{l-1}} \sum\limits_{\gamma \in A_l} c_{l-1}(\gamma) = 1$.

8.6.3 Sensitivity Analysis of the Neocognitron

The drawbacks of the original Neocognitron include the following: (1) Too much domain expert knowledge is required to train the Neocognitron; (2) there is no systematic way to specify the parameters; (3) it is unconvincing that the training patterns have included all the cases of the deformation and position shift. In this section, sensitivity analysis will play the role of analyzing the parameters for Neocognitron.

The response of the network to weight and input errors determines the sensitivity of the neural network. Being a visual pattern recognition system, the Neocognitron responds to input patterns by deciding which category the input belongs to. For the study of sensitivity, the response of the network to weight and input errors is called

STEP 1. Let δW be the weight perturbation ratio, N be the total number of stages of the Neocognitron, ξ_l, A_{S_l} be the selectivity and the receptive field of S-cell at stage l.

STEP 2. Let $l = 1$.

STEP 3. Calculate $\beta_l = \cos^{-1}[\xi_l/(1+\xi_l)]$.

STEP 4. Calculate the expected decision error at stage 1:

$$E(D.E.)_1 = 2f((\cos^{-1}(\cos\delta X \cdot \cos\delta W)/2), \beta_l, A_{S_l}).$$

where $f(.)$ is defined in according to Eq. (7.4).

STEP 5. For $l = 2$ to N :

$$\theta_X = E(D.E.)_{l-1} / f(\beta_{l-1}, \beta_{l-1}, A_{S_{l-1}}).$$

$$E(D.E.)_l = 2f((\cos^{-1}(\cos\theta_X \cdot \cos\delta W)/2), \beta_l, A_{S_l}).$$

STEP 6. $E(D.E.)_N$ is the computed sensitivity of the Neocognitron.

Fig. 8.4 The algorithm of sensitivity calculation for the Neocognitron

decision error, which is defined as the absolute difference between the decision made by the reference system and that made by the perturbed system. In Chap. 2, it has been mentioned that Oh and Marks (1989) and Stevenson, Winter and Widrow (1990) use the probability of decision error to represent the sensitivity. However, it is not suitable for the Neocognitron because the output of the transfer function of the Neocognitron is not a continuous one but rather a discrete one. Therefore, we consider not only the probability of decision error but also the magnitude of the error. Instead of a probability technique, the expected decision error is used to represent the sensitivity of the Neocognitron. The input vectors and the dimensional weight vectors of the reference system are called the *original input vectors* and the *original weight vectors,* respectively. The decision error is determined by the reference system and the perturbed system. The weight vector of the perturbed system is called the *perturbed weight vector.* In using the activation function, it is intuitively believed that the decision error is determined by the original weight vector, the perturbed weight vector, and the input vector.

In order to provide a clear picture in determining the sensitivity of the Neocognitron, a complete algorithm for calculating the sensitivity of the Neocognitron is presented in Fig. 8.4, where the function $f(a,b,n)$ is defined as follows.

$$f(a,b,n) \triangleq \int_0^a \frac{K_{n-1}}{K_n}\left(\frac{\cos u - \cos b}{1-\cos b}\right)\left(\sin^{n-2} u\right) du$$

$$+2\int_a^b \int_0^{\sin^{-1}} (\tan a \cot u \frac{K_{n-2}}{K_n}\left(\frac{\cos u - \cos b}{1-\cos b}\right)\left(\sin^{n-2} u\right)\left(\cos^{n-3} v\right) dvdu$$

where $K_n = 2\pi^{n/2} / \Gamma(n/2)$ and $\Gamma(\cdot)$ is the gamma function.

Based on the above sensitivity analysis, Cheng and Yeung (1999) concluded that the Neocognitron is sensitive to its receptive field, selectivity and the number of planes. Accordingly, Shi, Dong and Yeung (1999) proposed an evolutionary way to tune the parameters of the Neocognitron.

Bibliography

Akaike, H. (1974). A new look at the statistical model identiĉation. *IEEE Transactions on Automatic Control*, 19(6):716–723.

Alippi, C., Piuri, V. and Sami, M. (1995). Sensitivity to errors in artificial neural networks: a behavioral approach, *IEEE Transactions on Circuits and Systems – I: Fundamental Theory and Applications*, 42(6):358–361.

Anthony, M., Bartlett, P. L. (1999). *Neural Network Learning: Theoretical Foundations*, Cambridge University Press.

Battiti, R. (1994). Using mutual information for selecting features in supervised neural net learning, *IEEE Trans. on Neural Networks*, 5(4):537–550.

Bianchini, M., Frasconi, P. and Gori, M. (1995). Learning without local minima in radial basis function networks, *IEEE Transactions on Neural Networks*, 6(3):749–756.

Bishop, C. M. (1991). Improving the generalization properties of radial basis function neural networks, *Neural Computation*, 3(4):579–581.

Basson, E., Engelbrecht, A. P. (1999). Approximation of a function and its derivatives in feedforward neural networks, In: *International Joint Conference on Neural Networks*, Wasington, DC.

Bishop, C. M. (1995). Training with noise is equivalent to Tikhonov regularization, *Neural Computation*, 7(1):108–116.

Bishop, C. M. (1995). *Neural Networks for Pattern Recognition*. Oxford, U.K.: Clarendon.

Blake, C. L. and Merz, C. J. (1998). UCI Repository of machine learning databases. *School of Information and Computer Science, University of California, Irvine, CA*. [Online]. Available: http://www.ics.uci.edu/~mlearn/MLRepository.html.

Cao, X. R. (1985). Convergence of parameter sensitivity estimates in a stochastic experiment, *IEEE Transactions on Automatic Control*, 30(9):845–853.

Chakraborty, D., Pal, N. R., (2001). Integrated feature analysis and fuzzy rule-based system identification in a neuro-fuzzy paradigm. *IEEE Transactions on Systems, Man, and Cybernetics, Part B*, 31(3): 391–400.

Chen, S., Crown, C. F. and Grant, P. M. (1991). Orthogonal least squares learning algorithms for radial basis function networks, *IEEE Transactions on Neural Networks*, 2(2):302–309.

Chen, S., Chng, E. S. and Alkadhimi, K. (1996). Regularized orthogonal least squares algorithm for constructing radial basis function networks, *International Journal of Control*, 64(5):829–837.

Chen, S., Wu, Y. and Luk, B. L. (1999). Combined genetic algorithm optimization and regularized orthogonal least squares learning for radial basis function networks, *IEEE Transactions on Neural Networks*, 10(5):1239–1243.

Cheng, A. Y. and Yeung, D. S. (1999). Sensitivity analysis of neocognitron, *IEEE Transactions on System, Man, and Cybernetics—Part C: Applications and Reviews, 29*(2): 238–249.

Choi, J. Y. and Choi, C. H. (1992). Sensitivity analysis of multilayer Perceptron with differentiable activation functions, *IEEE Transactions on Neural Networks*, 3(1):101–107, 1992.

D.S. Yeung et al., *Sensitivity Analysis for Neural Networks*, Natural Computing Series, DOI 10.1007/978-3-642-02532-7, © Springer-Verlag Berlin Heidelberg 2010

Cloete, I. (1996). An algorithm for fusion of rules and artificial neural networks. In: *Proceedings of Workshop on Foundations of Information/Decision Fusion: Applications to Engineering Problems*, 40–45.

Cloete, I. (2000a). VL1ANN : Transformation of rules to artificial neural networks. In: *Knowledge-Based Neurocomputing*, MIT Press.

Cloete, I., Snyders, S., Yeung, D. S. and Wang, X. (2004). Sensitivity analysis of prior knowledge in knowledge-based neurocomputing. In: *Proceedings of 2004 International Conference on Machine Learning and Cybernetics*, 7:4174–4181.

Cloete, I. and Zurada, J. M. (2000b). *Knowledge-Based Neurocomputing*. MIT Press.

Davis, G. W. (1989). Sensitivity analysis in neural net solution, *IEEE Transactions on Systems, Man, and Cybernetics*, 19(5):1078–1082.

Dündar, G. and Rose, K. (1995). The effects of quantization on multilayer neural networks, *IEEE Transactions on Neural Networks*, 6(6):1446–1451.

Engelbrecht, A. P. (1999). Sensitivity analysis of multilayer neural networks, Ph.D. thesis, *The University of Stellenbosch*, South Africa.

Engelbrecht, A. P. and Cloete, I. (1999). Incremental learning using sensitivity analysis. In: *Proceedings of IEEE International Joint Conference on Neural Networks*, 2:1350–1355.

Engelbrecht, A. P., Cloete, I. and Zurada, J. M. (1995). Determining the significance of input parameters using sensitivity analysis. In: *Proceedings of the International Workshop on Artificial Neural Networks*, 382–388.

Engelbrecht, A. P., Fletcher, L. and Cloete, I. (1999). Variance analysis of sensitivity information for pruning multilayer feedforward neural networks. In: *Proceedings of IEEE International Joint Conference on Neural Networks*, 3:1829–1833.

Fu, L. and Chen, T. (1993). Sensitivity analysis for input vector in multilayer feedforward neural networks. In: *Proceedings of IEEE International Conference on Neural Networks*, 1: 215–218.

Fukuda, T. and Shibata, T. (1992). Theory and applications of neural networks for industrial control systems, *IEEE Transactions on Industrial Electronics*, 39(6): 472–489.

Gallant, S. I. (1993). *Neural Network Learning and Expert Systems*, MIT Press.

Gao, Z. and Uhrig, R. E. (1992). Nuclear power plant performance study by using neural networks, *IEEE Transactions on Nuclear Science*, 39(4):915–918.

Glanz, F. H. (1965). *Statistical extrapolation in certain adaptive pattern recognition systems*. Ph.D. thesis, Stanford Electronics Labs., Stanford, CA, Technical Report 6767–1.

Hagan, M. T., Demuth, H. B. and Beale, M. (1996). *Neural Network Design*. PWS Publishing.

Hashem, S. (1992). Sensitivity analysis for feedforward artificial neural networks with differentiable activation functions. In: *Proceedings of International Joint Conference on Neural Networks*, 1:419–424.

Haykin, S. (1994). *Neural Networks: A Comprehensive Foundation*. Macmillan College Publishing, New York, NY.

Haykin, S. (1999). *Neural Networks*, 2nd Edition, Prentice Hall.

Hoff, M. E., Jr., (1962). *Learning phenomena in networks of adaptive switching circuits*. Ph.D. thesis, Stanford Electronics Labs., Stanford, CA, Technical Report 1554–1.

Holtzman, J. M. (1992). On using perturbation analysis to do sensitivity analysis: derivatives versus differences, *IEEE Transactions on Automatic Control*, 37(2):243–247.

Hwang, Y. S. and Bang, S. Y. (1997). An efficient method to construct a radial basis function neural network classifier, *Neural Networks*, 10(9):1495–1503.

Hsu, C. W. and Lin, C. J. (2002). A comparison of methods for multi-class support vector machines, *IEEE Transactions on Neural Networks*, 13(2):415–425.

Huang, G., Liang, N., Rong, H., Saratchandran, P., Sundarasimhn, N. (2005). On-Line Sequential Extreme Learning Machine. *Computational Intelligence 2005*: 232–237.

Hyvärinen, A. and Oja, E., (2000). Independent component analysis: algorithms and applications, *Neural Networks*, 13(4-5):411–430.

Jabri, M. and Flower, B. (1991). Weight perturbation: an optimal architecture and learning technique for analog VLSI feedforward and recurrent multilayer networks, *Neural Computation* 3(4):546–565.

Jolliffe, I. T., (1986). *Principal Component Analysis*, Springer-Verlag, New York.

Kadirkamanathan, V. and Niranjan, M. (1993). A function estimation approach to sequential learning with neural networks, *Neural Computation*, 5(6):954–975.

Koda, M. (1995). Stochastic sensitivity analysis method for neural network learning, *International Journal of System Science*, 26(3):703–711.

Koda, M. (1997). Neural network learning based on stochastic sensitivity analysis, *IEEE Transactions on Systems, Man and Cybernetics – Part B: Cybernetics*, 27(1):132–135.

Kwak, N. and Choi, C.-H. (2002). Input Feature Selection for Classification Problems, *IEEE Transactions on Neural Networks*, 13(1):143–159.

Lacher, R. C., Hruska, S. I. and Kuncicky, D.C. (1992). Backpropagation learning in expert networks, *IEEE Transactions on Neural Networks*, 3(1):62–72.

Lamers, M. H., Kok, J.t N. and Lebret, E. (1998). A multilevel nonlinearity study design, In: IEEE *International World Congress on Computational Intelligence, International Joint Conference on Neural Networks*.

Lee, Y. and Oh, S. H. (1994). Input noise immunity of multilayer Perceptrons. *ETRI Journal*, 16(1):35–43.

Lovell, D., Bartlett, P. and Downs, T. (1992). Error and variance bounds on sigmoidal neurons with weight and input errors, *Electronic Letters*, 28(4):760–762.

Mao, K. Z. (2002). RBF neural network center selection based on Fisher ratio class separability measure, *IEEE Transactions on Neural Networks*, 13(5):1211–1217.

Michie, D., Spiegelhalter, D. J. and Taylor, C. C. (1994). Machine learning, neural and statistical classification. [Online]. Available: http://www.liacc.up.pt/ML/statlog/datasets.html.

Minsky, M. and Papert, S. (1969). *Perceptrons*. MIT Press, Cambridge.

Moody, J. and Darken, C. J. (1989). Fast learning in networks of locally-tuned processing units, *Neural Computation*, 1(2):281–294.

Murphy, P. M. and Aha, D. W. (1994). UCI repository of machine learning databases. *Department of Information and Computer Science*, Irvine, CA: University of California.

Ng, W. W. Y. and Yeung, D. S. (2002). Input dimensionality reduction for radial basis neural network classification problems using sensitivity measure, In: *Proceedings of the First International Conference on Machine Learning and Cybernetics*, Xian, China.

Ng, W.W.Y., Dorado, A., Yeung, D.S., Pedrycz, W. and Izquierdo, E. (2007). Image classification with the use of radial basis function neural networks and the minimization of localized generalization error, *Pattern Recognition* 40(1):19–32.

Oh, S. and Marks II, R. J. (1989). Noise sensitivity of projection neural networks, In: *Proceedings of IEEE International Symposium on Circuits System*, 3:2201–2204.

Oh, S. -H. and Lee, Y. (1995) Sensitivity analysis of single hidden-layer neural networks with threshold functions, *IEEE Transactions on Neural Networks*, 6(4):1005–1007.

Omlin, C. W. and Snyders, S. (2003). Inductive bias strength in knowledge-based neural networks: application to magnetic resonance spectroscopy of breast tissues, *Artificial Intelligence in Medicine*, 28(2):121–140.

Orr, M. J. L. (1995). Regularization in the selection of radial basis function centers, *Neural Computation*, 7(3):606–623.

Park, H., Murata, N. and Amari, S. I. (2004). Improving generalization performance of natural gradient learning using optimized regularization by NIC, *Neural Computation*, 16(2): 355–382.

Park, J. and Sandberg, I. W. (1993). Approximation and radial basis function networks, *Neural Computation*, 5(2):305–316.

Panchapakesan, C., Palaniswami, M., Ralph, D. and Manzie, C. (2002). Effects of moving the centers in an RBF network, *IEEE Transactions on Neural Networks*, 13(6):1299–1307.

Piché, S. W. (1992). *Selection of weight accuracies for neural networks*. Ph.D. thesis, Stanford University, Stanford, CA.

Piché, S. W. (1995). The selection of weight accuracies for Madalines, *IEEE Transactions on Neural Networks*, 6(2):432–445.

Poggio, T. and Girosi, F. (1990). Networks for approximation and learning. In: *Proceedings of the IEEE*, 78(9):1481–1497.

Principe, J. C., Euliano, N. R. and Lefebvre, W. C. (1999). *Neural and Adaptive Systems*, John Wiley & Sons, Interactive Electronic Book.

Quinlan, J. R. (1993). *C4.5: Programs for Machine Learning*. Morgan Kaufmann, San Mateo, CA.

Rosenblatt, F. (1958). The perceptron: a probabilistic model for information storage and organization in the brain. *Psychological Review*, 65:386–408.

Rumelhart, D.E., Hinton, G.E. and Williams, R.J. (1986). Learning internal representations by error propagation. In McClelland, J.L. and Rumelhart, D.E. (Eds.) *Parallel distributed processing*, Vol. 1. (pp. 318–362). Cambridge, MA: MIT Press

Schölkopf, B., Sung, K. K., Burges, C. J. C., Girosi, F., Niyogi, P., Poggio, T. and Vapnik, V. (1997). Comparing support vector machines with Gaussian kernels to radial basis function classifiers, *IEEE Transactions on Signal Processing*, 45(11):2758–2765.

Shi, D., Yeung, D. S., Gao, J. (2005). Sensitivity analysis applied to the construction of radial basis function network, *Neural Networks*, 18(7):951–957.

Stevenson, M., Winter, R. and Widrow, B. (1990). Sensitivity of feedforward neural networks to weight errors, *IEEE Transactions on Neural Networks*, 1(1):71–80.

Stevenson, M. (1990). *Sensitivity of Madalines to weight errors*. Ph.D. thesis, Stanford University, Stanford, CA.

Snyders, S. and Omlin, C.W. (2000). What inductive bias gives good neural network training performance. In: *Proceedings of the IEEE-INNS-ENNS International Joint Conference on Neural Networks*, 3:445–450.

Towell, G. G. and Shavlik, J. W. (1994). Knowledge-based artificial neural networks. *Artificial Intelligence*, 70(1-2):119–165.

Vapnik, V. (1995). *The Nature of Statistical Learning Theory*, Springer.

Wang, Z. and Zhu, T. (2000). An efficient learning algorithm for improving generalization performance of radial basis function neural networks, *Neural Networks*, 13(4):545–553.

Widrow, B. and Hoff, M.E. Jr., (1960) Adaptive switching circuits, *IRE Western Electric Show and Convention Record*, 4:96–104.

Winter, R. G. (1989). *Madaline rule II: a new method for training networks of Adalines*. Ph.D. thesis, Stanford University, Stanford, CA.

Wright, W. (1999). Bayesian approach to neural network modeling with input uncertainty. *IEEE Transactions on Neural Networks*, 10(6):1261–1270.

Xu, L., Krzyzak, A. and Yuille, A. L. (1994). On radial basis function nets and kernel regression: statistical consistency, convergence rates and receptive field size, *Neural Networks*, 7(4): 609–628.

Yeung, D. S. and Wang, X. Z. (1999). Initial analysis on sensitivity of multilayer Perceptron, In: *Proceedings of IEEE International Conference on Systems, Man and Cybernetics*, 3:407–411.

Yeung, D. S. and Sun, X. (2002). Using function approximation to analyze the sensitivity of MLP with antisymmetric squashing activation function, *IEEE Transactions on Neural Networks*, 13(1):34–44.

Yeung, D. S., Ng, W.W.Y., Wang, D., Tsang, E.C.C. Tsang and Wang, X-Z. (2007). Localized generalization error and its application to architecture selection for radial basis function neural network, *IEEE Transactions on Neural Networks* 18(5):1294–1305.

Zeng, X. and Yeung, D. S. (2001). Sensitivity analysis of multilayer Perceptron to input and weight perturbation, *IEEE Transactions on Neural Network*, 12(6):1358–1366.

Zeng, X. and Yeung, D. S. (2003). A quantified sensitivity measure for multilayer Perceptron to input perturbation, *Neural Computation*, 15:183–212.

Zurada, J. M., Malinowski, A. and Cloete, I. (1994). Sensitivity analysis for minimization of input data dimension for feedforward neural networks. *IEEE International Symposium on Circuits and Systems*, 6:447–450.

Zurada, J. M., Malinowski, A. and Usui, S. (1997). Perturbation method for deleting redundant inputs of Perceptron networks, *Neurocomputing*, 14(2):177–193.